POPULATION, ECONOMIC DEVELOPMENT, AND THE ENVIRONMENT

EDITED BY

Kerstin Lindahl–Kiessling
and
Hans Landberg

OXFORD UNIVERSITY PRESS
1994

Oxford University Press, Walton Street, Oxford OX2 6DP
Oxford New York
Athens Auckland Bangkok Bombay
Calcutta Cape Town Dar es Salaam Delhi
Florence Hong Kong Istanbul Karachi
Kuala Lumpur Madras Madrid Melbourne
Mexico City Nairobi Paris Singapore
Taipei Tokyo Toronto
and associated companies in
Berlin Ibadan

Oxford is a trade mark of Oxford University Press

Published in the United States
by Oxford University Press Inc., New York

British Library Cataloguing in Publication Data
Data available

Library of Congress Cataloging in Publication Data
Data available
ISBN 0–19–828950–2

1 3 5 7 9 10 8 6 4 2

Printed in Great Britain
on acid-free paper by
Biddles Ltd.,
Guildford & King's Lynn

Preface

The chapters that make up this volume have sprung from the conviction that we need a deeper understanding of the complex interactions between the social systems and the physical, chemical, and biological interactive processes that regulate the earth system and, consequently, also provide the unique environment that makes human life as we know it today possible. Changes and variations occur incessantly in these interactive processes and, thus, the way in which human activities influence them will interfere with the functioning of the whole global ecosystem.

The interlinkages between humankind and the natural environment have always been obvious to the common people and they have been analysed and discussed in scientific terms for centuries. However, the industrial and scientific revolutions provided new opportunities for the exploitation of the natural resource base to a degree that seemed infinite. In addition, the ability of human ingenuity to cope with the environmental disturbances created by humans was conceived as guaranteed.

The rapid increase in consumption levels in the wealthy regions of the world and the rapid growth in world population—with the haves eager at least to preserve what they have gained, and the have-nots, with all good reason, claiming an equal share in the increasing standard of living or 'good life'—have, during the last decades, made it obvious to almost everyone that natural resources are fragile and the resilience of the world's ecosystems is limited. The limited capacity of change of the basic socio-economic systems and the population growth create, together with these limitations, a real threat to mankind and man as part of nature. It is not a case of 'us' and 'them'—time is running short for all of us. The environmental impact of high consumption levels in the industrialized world endows the rich countries with a problem of no less importance than fast population growth in the South. It would seem that for scientists and politicians alike the message is clear: the issues of population, development, natural resources, and environment must be considered together.

Against this background it became a matter of great concern when, in

1991, representatives of the Royal Swedish Academy of Sciences (KVA) and the Swedish Council for Planning and Co-ordination of Research (FRN), who were members of the Swedish National Preparatory Committee for the UN Conference on Environment and Development, discovered at the outset of the UNCED planning process that population issues had not been included on the UNCED agenda. Given the time-frame available, it was decided that a realistic contribution could be made by organizing a conference that would engage scientists from a wide range of disciplines. The purpose was to induce economists, ecologists, sociologists, demographers. and others to meet on an equal footing in order to present to political audiences a comprehensive analysis and a set of recommendations that addressed the relevant issues and sought to elucidate the pertinent linkages between them. The outcome of the deliberations was a widely distributed Conference Statement and a special issue of *Ambio* (21/1 (1992)).

Our concern about the linkages between population, environment, and development was shared in many quarters, and several contributions to the discussion on these issues led to their being at least formally considered during the deliberations of the Earth Summit Conference in Rio de Janeiro in June 1992. Outside the official conference many heads of states expressed their awareness of the fundamental complementary relations between socio-economic development and the life-supporting environment, most clearly voiced by Prime Minister Gro Harlem Brundtland, Norway: 'Poverty, environment and population can no longer be dealt with—or even thought of—as separate issues; they are interlinked in practice and cannot be de-linked in the formulation of policies.'

In part the lack of real interest in population issues at UNCED can be explained by the fact that a new UN Conference on Population and Development was already planned and scheduled for Cairo in 1994. This in itself is a good illustration of the more-and-more obsolete sectorization which impedes coherent international and national policy-making, as well as science. The task this time must be firmly to place the environmental concerns on the Cairo agenda together with the main topic: *the human population and its future*, because, after all, population, development, natural resources, and environment are undoubtedly one and the same issue. This has already been symbolically demonstrated at the 1972 UN Conference on the Human Environment in the Conference concept *Only One Earth.*

Consequently, KVA and FRN, now in co-operation with the Beijer International Institute of Ecological Economics, are endeavouring to inject an environmental perspective into the UN Cairo Conference. The material presented in this volume is based on a series of seminars at which seven distinguished scholars from different fields of study delivered public lectures, the aim of which was to illuminate the population, environment, development nexus from a number of interrelated perspectives. All authors present their own view of the central feature of this complex issue, which is not always in agreement with our earlier document. Special reference is also made to the *Ambio* issue referred to above, which will provide the reader with a number of other relevant contributions on this subject.

Each lecture was followed by a discussion among the lecturers and a reference group of around twenty Scandinavian scholars, who were specially invited to participate in the entire series. Three overriding themes were: the Malthusian Conflict; Factors underlying Fertility Changes; and Strategic Development Issues Related to the Population–Environment Nexus. A final round-table meeting was held on 7–8 January 1994, based on the material presented during the whole series. In this book the reader is invited to take part in the intellectual outcome of this work. The lectures, presented in Chapters 3–9, are preceded by an introduction and two chapters that build from, and summarize, both the lectures and the ensuing seminar discussions. The introduction is jointly authored by the steering committee, Chapters 1 and 2 by members of the steering committee in collaboration with members of the reference group.

Obviously, it would be naïve and pretentious to believe that international scientific or even political conferences can, in themselves, achieve more than raising human consciousness and perhaps pointing to and/or taking a few stumbling steps in the right direction. The responsibility of the scientists is, however, obvious: to put the questions right, to analyse the alternatives and consequences, to demonstrate the interlinkages, and to create the integrative knowledge. However, final responsibility for taking action rests with politicians. Their task is indeed difficult: not only must they judge between scientific opinions all too often supporting quite different or contradicting strategies, they must do so under the pressure from what might be their own conflicting goals and the seemingly unavoidable economic competition between nations and regions.

The inclination to postpone, until a disaster is already knocking at the door, any action which involves sacrificing our current standard of living in order to escape the risk of future catastrophes is manifest in all human societies. The resilience of the ecological system has even compounded this strategy. Many warnings of environmental breakdowns have been exaggerated or, as seen in retrospect, simply false. It is worth considering, however, that the rapidly increasing stress on our natural-resource base can, especially in the overpopulated areas of the world, create social tension and conflicts between or within nations long before major ecological breakdown occurs. The examples are there already. There is no time to lose. It may be in all our interests to consider Kenneth Boulding's reflection (in R. Constanza's *Ecological Economics*) with true thoughtfulness:

Those who are concerned with the future of the planet in its totality have to ask themselves: What are the sources of power in the human species directed towards changing the future in a desirable direction? It is not enough to preach 'we must do this, we must do that'. Such preaching is perhaps a necessary prerequisite to action, but it does not necessarily produce the action required . . . We do need to study very carefully, therefore, the structure of power in the human race over the future, which involves not only particular decisions but also the overall images of the world in the minds of the decision makers, and the learning process by which these images are created . . . And in the solutions of these problems—if there is one—we will undoubtedly have to make some use of threat power—taxation, regulations and so on—and some use of economic power. But the major element unquestionably will be integrative power, based first on widespread knowledge that we all live on the same fragile planet, which we have now seen from outer space and to which we owe a common loyalty and affection. Unless this view is very widespread, legitimacy will not be granted to those frequently painful processes which may be necessary to prevent catastrophe.

The key to progress is hidden in the ability to integrate, the ability not only to see but also to understand that we have indeed only one fragile Earth.

Kerstin Lindahl–Kiessling
Hans Landberg

Steering Committee

Professor Partha Dasgupta (president)
University of Cambridge

Professor Kerstin Lindahl-Kiessling (vice-president)
The Royal Swedish Academy of Science

Professor Hans Landberg (vice-president)
The Swedish Council for Planning and Coordination of Research

Professor Erik Arrhenius
University of Stockholm

Assoc. Professor Tommy Bengtsson
Lund University

Professor Karl-Göran Mäler
The Beijer Institute, the Royal Swedish Academy of Sciences

Professor Claes Ramel
University of Stockholm and the Royal Swedish Academy of Sciences

Assoc. Professor Uno Svedin
The Swedish Council for Planning and Co-ordination of Research

Professor Sture Öberg (Scientific Secretary)
Uppsala University

Acknowledgements

Our sincere and warm thanks go to the authors for their contributions and for their willingness to participate in this somewhat unorthodox step-by-step conference. Special recognition goes also to Dr Bo Malmberg, who, together with Professor Sture Öberg, was responsible for the scientific secretariat to the conference. The main burden of recording and summarizing the discussions and putting the manuscripts together was carried out by Dr Malmberg. Efficient administration was carried out during the whole process by Christina Leijonhufvud and her colleagues at the Beijer Institute.

Funding for the project from the Royal Swedish Academy of Science, the Swedish Council for Planning and Co-ordination of Research, and the Beijer International Institute of Ecological Economics is gratefully acknowledged.

Contents

1. Population, Development, and Institutional
   Change: Summary and Analysis
   Tommy Bengtsson and Christer Gunnarsson   1

2. The Environmental Resource Base and Human Welfare
   Partha Dasgupta, Carl Folke, and Karl-Göran Mäler   25

3. Population and Reasoned Agency: Food, Fertility,
   and Economic Development
   Amartya Sen   51

4. An Ecologist View of the Malthusian Conflict
   C. S. Holling   79

5. 'Children are like young bamboo trees':
   Potentiality and Reproduction in Sub-Saharan Africa
   Caroline Bledsoe   105

6. Economic Analysis of Fertility: Micro-Foundations
   and Aggregate Implications
   Robert J. Willis   139

7. Government, Population, and Poverty: A 'Win-Win' Tale
   Nancy Birdsall   173

8. Institutional Analysis of Fertility
   Geoffrey McNicoll   199

9. The Relevance of Malthus for the Study of
   Mortality Today: Long-Run Influences on
   Health, Mortality, Labour Force Participation,
   and Population Growth
   Robert Fogel   231

Introduction

SETTING THE SCENE

Public concern over the issues of population growth and the sustainability of the natural environment and resource base has a long tradition. In the accompanying scientific debate, discrimination between essential problems and unnecessary anxieties has not always proved to be easy; the old Malthusian conflict remains unresolved. Questions of technological development, societal adaptation, the environmental limits to growth, and the concept of discontinuities and irreversible ecological collapse are the subjects of considerable scientific controversy.

For this reason, this book approaches the complex issues related to the population–environment–development nexus through the Malthusian conflict. Malthus' basic question was: will there be enough food for everyone? 200 years later, global population increase and production and consumption patterns, in conjunction with their unequal distribution between the poor and the rich, and the limits to productivity of ecological systems make the question as valid as ever.

The approach to the problem and the scope of the debate are characterised by the position, strongly maintained in this book, that issues of population growth or decline cannot be separated from questions of economic and social development, or from the environmental concerns related to the management of resources on a global level.

One neglected issue is the question of the resilience of social systems faced with stress emanating from environmental degradation or increasing scarcity of essential natural resources. There is no simple relationship between economic well-being and social adaptability towards environmental change. Although technologically sophisticated societies seem more vulnerable in many respects than less-developed countries, the latter will in most cases be harder hit by competition for natural resources. There is good reason to be aware of the many possible vicious circles where environmental stress creates societal disturbances and which in turn increase environmental problems or impede development.

These concerns underline the need to analyze institutional issues, and the latter have come to the forefront in development economics, contributing strongly to broadening the discussion of development problems and to opening up and improving the possibilities for integration and collaboration between economics and other social sciences as well as ecology. This discourse is the foundation for the second and third theme of this book, illustrating that a coherent theoretical framework is slowly evolving in the field of population development environment research.

Poverty is one aspect of the development complex on which considerable attention has been focused. The analysis here concentrates on crucial questions of conditions and behaviour of rural households as one of the important springs for the relevant actions in this social interaction process. Environmental issues are central to the complex of problems. Advanced ecological systems research underlines that the productivity of natural systems, in spite of their often surprising resilience, does have limits and that signs of stress on the global environment are evident. Adaptations and changes occur all the time due to inherent incidents within the ecological systems or due to events of geological character but also as 'responses' to human activities. The scale and scope of economic activity and consumption demands of a rapidly increasing world population now heavily increase stress on the environment and build up also for sudden and unexpected major changes. They, in turn, might give rise to series of other events before a new phase of relative stability is established. The current environmental changes on a smaller scale are in part possible to foresee and also more easily compensated for by individuals and societies. The major events are difficult or impossible to forecast, both in time, location, and magnitude, and are correspondingly more threatening. They may demand a capacity for adaptation which societies are unable to mobilize. Two general conclusions emerge from the discussion here. The first is the so called precautionary principle: even with incomplete scientific knowledge, new or increased burdens on the environment should be avoided as far as choices are available to us. However, existing and foreseeable burdens on nature also require urgent positive action in order to alleviate or at least prevent their acceleration. The avoidance or postponement of major collapses must be seen as a possibility, but the time-scale in which this can be done is narrowing.

MARKET PRICES AND SOCIAL VALUE

In the early economic literature on the environment, failures in the functioning of markets were often identified as the underlying cause of environmental problems. In connection with an increased interest in institutional questions in general recent research has however illustrated that certain patterns of environmental deterioration can be traced to inappropriate government policies, and not to market failures. The notion that markets, functioning only on their own terms, can take care of environmental problems has lost its validity. The institutional failures must be considered in a wider perspective.

One important example of this institutional failure, where governments have to act, is that the market prices of environmental resources do not reflect their social value. Many reasons why they possess worth have been presented but these reasons are not always appreciated or accepted, partly because they can represent an implicit threat to certain economic or social structures.

In economic terms the social worth of a commodity can be referred to as its accounting price (or shadow price). For environmental resources *in situ*, the gap between market and accounting prices are typically large and especially so in developing countries where private property rights to such resources are often absent. It means that their market prices are zero, even while their accounting prices are positive.

When market prices do not correspond to the social value of resources, but understate them, the economic system will, through the inappropriate incentives, overexploit these resources compared to what would have been socially optimal.

This is the basic reason why it is so important for poor countries to get their prices right. Furthermore, if the prices represent social values in a better way, the burden from population pressure will be reduced and on the other hand demonstrate the true costs for population growth.

How can prices be corrected, where prices are at all possible to establish? By a variety of means, although it must be realized that there are few easy or politically comfortable ways. By defining property rights, where that can be done, market prices will tend to respond in a better way to social values. It must be stressed, however, that such a recommendation does not necessarily imply private property rights: there may be times when collective ownership or stewardship may be preferred. The important thing is that someone, or some well defined group, can be assured of the benefits of socially desirable management and also be held accountable for mismanagement. If, for various reasons, it proves impossible to define property rights, the government can by taxes or subsidies change the market prices so as to make them correspond better to the social values. In some cases it may be necessary to impose regulations in order to bring about the desired uses. In these cases, market prices will adjust to the implicit change in the supply of the resources and thereby better reflect their social value.

The whole matter of accounting prices versus market prices is closely related to the manner in which real net national product (NNP) should be measured. It is argued in this book that, if NNP is to reflect social well-being, accounting prices should be used in its estimation, not market prices. If this is done consistently, what is often referred to as the green NNP can be calculated. The estimation of accounting prices for environmental resources is thus a vital step for national economies.

POPULATION EXTERNALITY

There is another failure that is of crucial significance when considering fertility and population: the failure connected with fertility itself. As is shown in this book, there is no single factor underlying fertility changes. However, there are two broad sets of determinants: family-planning techniques available to households, and the demand for children by households. It is also clear that the demand for children depends on a number of economic conditions such as the cost of raising children, the return from children as labour on the household farm or from collecting fuel-wood, the need for old age security and so on. The demand for children will typically be determined only by factors that affect the household directly. If high population growth will increase the destruction of the environment, this factor will not be taken into account when decisions on the size of the household are being taken. Thus there is an externality connected with fertility decisions, an externality that is the focus of this book.

This externality exists everywhere, but it is strongest in the rich countries as a new individual in one of these will exploit more of the natural resource base than a new individual in a poor country will. In this sense, the externality generated by population growth in the North is much more serious. However, that is not an excuse for neglecting the corresponding externality in the South, as the people in the South can exert control themselves over that externality and thereby improve their situation.

Thus, the industrialized world needs a population programme as much as the developing countries. In general terms the North should reduce their excessive resource consumption, for example by appropriate pricing reforms, and the South has to implement policies in order to reduce their population growth as part of their programme for sustainable development. But no single measure will suffice. Typically, there seems to be a need for a mixture of policies that combine reproductive health services, including family planning, improved employment prospects for women, improved education and literacy programmes, and a more extensive availability of basic needs goods, such as potable water, household fuel, and credit savings and insurance facilities. In the design of such policies, it is imperative to take into due account a wide range of women's rights, including, in particular, their rights to reproductive choice. It bears emphasis that the wanted and acceptable mixture will vary across regions (e.g. paddy versus wheat) and cultures (e.g. nuclear versus extended families). A pleasing feature of this insight is that what are desirable population policies are also those that are directed against poverty and environmental degradation. It is however judicious to observe that we are still relatively ignorant about the quantitative significance of the various factors across regions and cultures. Empirical investigation into these issues, in conjunction with local capacity building and leadership, should have high social priority.

INSTITUTIONAL CONSTRAINTS

The technical solutions to the problems regarding calculation and measurement of accounting prices for environmental resources are now, albeit sometimes complicated, in place or well under way. The main problems now are basically political in nature. Changes of the kind touched upon above affect power structures at all levels in societies. One of their prerequisites and/or consequences is that new norms, rules and power relations have to be established, or old ones transformed. Without at all underestimating the difficulties related to institutions when, for example, a reduction in the rich world of its environmentally harmful use of energy becomes necessary, it must be stated that the problems are worse in some developing countries. It is especially so in countries facing environmental and economic pressure as a result of precipitate population growth. The urgency of these problems as they are manifested in the Indian sub-continent and Sub-Saharan Africa, land masses that are inhabited by some 2 billion people, is well demonstrated in this book. It is commonplace to say that people in these regions face severe constraints in their ability to engage in production and exchange and that the constraints are due to the institutional failures that prevent markets and the public infrastructure from functioning with anything like the efficiency with which they operate in today's rich nations. Civil wars and strifes are worsening the situation in all respects but are themselves only proximate causes.

Both theory and clear evidence suggest that these institutional failures, over and above the day to day constraints, also cause a great uncertainty to the household regarding its future well-being. Many of the activities of poor households are interlinked and aim, at the same time, at circumventing the limitations they face in their economic transactions as well as at reducing uncertainty. To give an often demonstrated example, children in poor households are not only ends in themselves, they are also means by which the household obtains an increase in security for the future. Children play an important role in providing their parents with labour today and support during their old age. Another, more acute, example observed in this volume is that women facing deteriorating economic circumstances have been known to bear children with different men and thus through polyandrous motherhood creating access to more than one resource network. This means that children substitute for other sources of support, such as well-functioning capital and insurance markets or government pensions plans.

The policy implications are obvious. A household can only obtain security if its individual members are guaranteed protection against force, fraud and theft, access to well-functioning insurance and credit markets, basic needs (e.g. food, potable water, primary and secondary education, public-health services and family planning resources) and employment opportunities. The social methods that are needed to establish a well functioning resource allocation mechanism to this end certainly vary between regions and from culture to culture. In some situations, for example, agrarian reform could be a necessary precondition for the emergence of a well-functioning resource allocation mechanism. In others it may not be the essential tool. Thus, there is no single solution to the security and development problems. Certain institutional frameworks are in all circumstances essential but they can also be substituted. The security concept can be conceived as a compass for policymakers in their search for institutional change supportive to a positive population–environment–development process.

HOUSEHOLDS AND INDIVIDUALS

A switch in focus between households and individuals is also important. What transpires inside the household is usually deemed a private matter in modern cultures. Perhaps for this reason, much economic and social inquiry takes the household as the unit of analysis. But households contain micro-institutions, with their attendant strengths and failings. In any event it is the individual who should be the basic focus of moral concern. No account of fertility matters would be satisfactory if it were not to peer inside the household. Issues concerning inequality between the genders, and between young adults and the old, stem not only from institutional failure in the public sphere, but also from the norms and social patterns regulating the life within families and households. Power relations within the household can be highly unequal. Research has evidently shown that illiterate adult women facing few opportunities outside the home are in a particularly vulnerable position. As would be expected, this translates itself into gender inequalities among children as well (e.g. inequalities in their nutritional status). Thus, poverty, high fertility, and degradation of the local environment resource base are not only interlinked, they are in turn linked to the extent of inequality between the genders. The single most important contribution to a change in this institutional pattern is increased education and vocational training for girls and women.

BREAKING VICIOUS CIRCLES

There are many possible vicious circles where environmnental stress can create societal disturbances which, in turn, increase environmental problems or impede development. Any development plan will, therefore, be subject to shifting conditions, always under pressure from conflicting goals where long-term perspectives are obscured by urgent short-term needs. Such a plan cannot be carried out from the top down only, but must be incorporated with bottom-up efforts. Processes need to be initiated that, while giving positive spin-off effects, will unite or reconcile the long-term development goals with the urgent political and moral requirements, creating the objectives and incentives that will make it possible for individuals, households, communities, and societies to break the vicious circles they are caught up in and thus launch a truly sustainable development plan.

1

Population, Development, and Institutional Change: Summary and Analysis

Tommy Bengtsson and Christer Gunnarsson[*]

The general links between population, environment, and development are dealt with elsewhere in this book. Here the interplay between population and development will be considered in greater detail, especially the significance of institutions for development. The problem is not going to be that global birth rates will never go down. This they will certainly do, for the average size of family in the world has already fallen from 6.0 in 1960 to below 3.5 in 1990. From all that we know about the decline and its causes so far, fertility will continue to fall where it has started to do so. The problem instead is that in the poorest parts of the world neither development nor the fertility transition has started yet. Thus we have to find a way that will bring economic and social development and a fertility transition to the poorest without damaging the environment. This demands a policy that mobilizes local social structures and a timing and sequencing of interventions as the development policy alters local institutions—a policy that recognizes the social and cultural differences between different nations and during different stages of development.

The chapter is divided into five sections. In Section 1 we discuss general links between population and development. Section 2 considers what distinguishes poor societies from rich. Section 3 discusses how the transition from poverty to prosperity was accomplished in various parts of the world and Section 4 looks at how institutional development contributed to this process. Finally, in Section 5, we discuss what we have learnt from the past.

* Associate Professors in Economic History at Lund University.

1. THE LINKS BETWEEN POPULATION AND DEVELOPMENT

The number of very poor people in the world has increased during the post-war period at the same time as their *proportion* has decreased. The majority of countries have experienced both economic and demographic developments characterized by a substantial rise in living standards. This development started in Western Europe, North America, and Japan during the second half of the nineteenth century, and in Latin America, India, and China in the 1960s and 1970s. A number of countries have entered upon their fertility transition during the 1980s, while others still lag behind. Thus, the timing of both the economic and the demographic transformations has varied considerably. There seems to be no given scheme of development that all countries must follow, although certain common features can be discerned.

In a broad sense, economic and demographic development can be said to go hand in hand. In a long-term perspective, the demographic transition and the rapid rise in population characterizing the actual change-over phase have coincided with a modernization of society. Accordingly, it is not by chance that the demographic transition occurred during the nineteenth and early twentieth centuries in Western Europe in conjunction with the transformation of agriculture and industry and not during the Renaissance or Middle Ages. In the same way, it is no coincidence either that a demographic transition in what are today called the newly industrialized countries took place during the post-war period and not earlier. In Chapter 6 Robert Willis discusses how economic growth leads to a diminution of the proportion employed in agriculture and thereby to an increase in the need for skilled labour. This, in turn, leads to families opting to have fewer children and to educate them for work in the modern sector. (The argument is the same in Oshima's analysis of East Asia: see Oshima 1992). It may be added here that agriculture too is modernized during such a transformation, which generates a growing need for skilled labour in this sector as well.

In an agrarian society children start to work at an early age. This is especially important in a society with a poorly developed infrastructure, in which the children help with household tasks such as fetching water and so forth (Dasgupta 1992). Grown-up children spell security for the mother if the father dies. They also signify security in old age. The costs of bringing up children are low. Education costs are trifling and the indirect cost of the time the mother spends on the children is low. Willis integrates theories of aggregate economic–demographic interaction and micro-level theories of family decisions about fertility, investment in human capital, and allocations of time to explain the causes and consequences of demographic change. Nancy Birdsall, in Chapter 7, discusses different policies aiming at changing direct and indirect costs of children. This is not to be regarded as meaning that demographic transformation is an automatic reaction to economic development. The links between development and demographic change are more complicated than that. There have been many countries where mortality began to fall before industry and agriculture underwent their transformations. In other countries, family size diminished before infant mortality fell and before their industrialization took place (see Fridlizius 1984; Schofield 1984; Perrenound 1984; see also Bengtsson 1992). Thus, there is not always a decade-to-decade connection between the demographic transition and the revolution of the economy even though they are linked in the long-term perspective. Before we examine the different patterns of development we must discuss what it is that differentiates societies that are still in a pre-development stage from the ones where development has started —what differentiates pre-industrial from industrial societies.

2. POOR SOCIETIES AND RICH SOCIETIES—WHAT IS THE DIFFERENCE?

The answer to the question is not as straightforward as it seems. Before discussing differences between poor and rich countries of today, we will turn to the differences in the past. Until about 1800 every long-term increase of Britain's population caused prices of food to rise while wages were held back (Wrigley and Schofield 1981; see also Bengtsson 1992). The value of land rose while that of labour power fell. Developments were probably similar in other parts of Western Europe and probably also in other parts of the world. The reason was that agricultural production was rising more slowly than population despite land reclamation and land reforms, and despite improvements in agricultural techniques. When population increased year after year, therefore, the standard of living fell for the vast majority. The changes in England in the early nineteenth century, the extensive clearing of new land, the land reforms, the new varieties of crop, the intensification of crop rotation, the improved manuring, and the new farm implements originating from the industrial sector—all these brought a revolution in agriculture. For the first time agriculture was able to supply the growing population with sufficient food. At the same time, the proportion of the population employed in agriculture decreased—more and more workers were finding jobs in the fast-growing industrial sector. For the first time the living standard of the vast majority rose simultaneously with an increase of population. England escaped from 'the Malthusian trap' and stepped forward into a society characterized by 'modern economic growth', as Simon Kuznets expressed it. Other countries of Western Europe and North America, and Japan, followed during the second half of the nineteenth century. Thus, differences in the degree to which population growth is associated with a rise or a fall in the standard of living is one way of distinguishing a poor pre-industrial from a rich industrial society.

Another way of describing the pre-industrial society is by drawing attention to its capacity for maintaining a long-term equilibrium between population growth and economic development. Long periods of population growth and falling living standards lead to either rising mortality, falling birth rates, or large-scale emigration. Malthus believed that mortality was the primary factor that prevented the population from growing beyond all bounds. This emerges most clearly in the first edition of his essay. Later he modified this conclusion, although he never abandoned it. Modern research has shown that in fact it was birth rates that fell, as a result not of reduced family size but of fewer marriages. This way of restoring equilibrium has been described as 'the nuptiality valve' (Wrigley and Schofield 1981). In other words, the existence of obstacles to population increase—increased mortality, decreasing family size or fewer marriages—may constitute criteria of a poor pre-industrial society.

A third possibility is to take mental note of the uncertainty that prevailed in the pre-industrial society. In that society there was great uncertainty with respect to the prospects of food supplies, as harvest outcomes were difficult to forecast. If the crops failed, almost a whole year of hunger followed. If the winter was unusually cold, fuel for heating would be insufficient. Hunger drove people out on the roads in search of work. Another element of uncertainty concerned the supply of land. Tenancies could be terminated and altered, and even a peasant-proprietor might lose his holding. Legal uncertainties, with respect to both property and other matters, were great. It was hard for ordinary folk to protect themselves against injustices perpetrated by the ruling classes of society. Vast uncertainty with regard to conditions during the forthcoming year caused people in pre-industrial societies to spend most of their time, and their investments, endeavouring to create a more secure existence. This could manifest itself in a number of ways. In agriculture, the system of strip cultivation was one way of protecting oneself against extremely poor harvests. The village moot was the institution which organized production and ensured justice between the cultivators. The other great element of uncertainty concerned security in old age. However, planning for old age usually had to yield precedence to the need to secure the food supply for the years immediately ahead. The Agrarian Revolution meant not only a rise in the average standard of living but also increased security. With their daily bread secured, farmers could plan for the future, and for their old age.

The characterization above applies to the pattern of historical development of today's industrial countries. Can today's poor countries be characterized as poor pre-industrial countries in any of these senses? What does completely differentiate today's poor countries from Western Europe in the middle of the nineteenth century, for example, is that most other countries are industrial. Certainly, when Germany or Sweden was industrializing, the Industrial Revolution had already taken place in England, but the major part of the world was still at the same level as Germany or Sweden itself. Developing countries today not only are part of a world where most other countries are industrialized, but they most often also have a modern sector themselves. It tends to be more usual than unusual for educational opportunities, hospitals, a developed banking system, and other such facilities to exist in the capitals of developing countries. But the vast majority of people in the poor countries live in the countryside, although urbanization is rapid in most developing countries, and the towns contain a large proportion of poor people as well. For most of them the modern sector, whether in their own country or abroad, constitutes a possibility at best. Another crucial difference is that mortality has declined sharply in the majority of today's poor countries without fertility having fallen to the same extent, with very rapid population growth as a result. Population growth is about three times more rapid than what it was during the demographic transition in today's industrial countries. Thus the existence of a modern world and the high population growth are examples of major differences between the poor countries in the past and the ones of today.

In the majority of the poorest countries the standard of living has been falling since the middle of the 1970s, while population has been growing at the same time. In this respect they can be said to be pre-industrial, not having attained the stage of 'modern economic growth'. As regards the insecurity factor, this is probably greater than it was in Western Europe immediately prior to the emergence of the modern society. Uncertainty applies not only to harvest results and world market prices but also to the economic and political system at home. One probably has to go back to the eighteenth century to find a similar era of uncertainty in Western Europe. These are very important characteristics that one has to bear in mind when discussing the development opportunities of the poor countries of today.

In one more respect, today's poorest countries differ from the way things were previously in today's industrial countries. Of course one cannot tar them all with the same brush, but in many cases, especially in Africa, there does not seem to be anything to prevent the population from growing rapidly, at least not in a short- or medium-term time perspective. This can be interpreted as either meaning that, when all is said and done, the conditions of life are not all that bad, or else that the factors that previously prevented the population from growing geometrically have been put out of action. Africa, of course, has not always had high population growth. During the nineteenth and early twentieth centuries it was instead the populations in Western Europe and North America that were growing fast. During the inter-war period and up to the beginning of the 1970s a number of African nations also seem to have enjoyed high economic growth and a steady increase in living standards, but since then the situation has deteriorated. In contrast to pre-industrial England, the birth rates seem to remain almost unchanged despite the falling standard of living. The explanation is that children contribute to the family's subsistence even when young, that they are regarded as providing the prospect of future income and security, and that the costs of children are shared by several families. Despite good availability of contraceptives, fertility remains high. Caroline Bledsoe, in Chapter 5, discusses this phenomenon in detail with examples from Sierra Leone, Liberia, and the Gambia. Here it also seems possible to form 'families' even in times of falling living standards. Fertility can, therefore, be said to be isolated from economics. This applies not only in a short- but also in a medium-term time perspective. Mortality too can be isolated from the economy, as shown by Robert Fogel in Chapter 9. Robert Fogel stresses that there is no equilibrium position between food supply and population numbers but that there are multiple equilibria. Undernourishment can arise in consequence of a food intake which is either too low or of excessively poor composition, or because the body requires more food than normal as a result of illness. A low food intake leads directly to death only in extreme cases and then usually in cases of nutrition-related illnesses (especially infectious diseases). This does not mean that low food intake does not have its consequences. On the contrary, the consequences are severe, especially if an individual suffers dietary deficiencies during the period when the body is being built up. Thus, protein calorie deficiency under age 3, for example, can lead to a permanent reduction in brain cells and to permanent stunting of the central nervous system. Again, even temporary disturbances of the food intake can lead to the failure of vital organs to develop as they should and to the individual's not having the length of life-span he otherwise would have enjoyed. The periods of infancy and adolescence are particularly sensitive, as is the foetus stage, but really the entire growing-up period is sensitive.[1] Thus, the effects of undernourishment are that the vital bodily organs do not develop fully, that the body's growth stops, and that the physical and mental work capabilities are lower. If food intake is very low, overall energy does not suffice for work but only for maintenance of the vital functions.

Because the body does not develop as it should, the life-span also becomes shorter than it would otherwise have been. This need not necessarily mean that mortality increases forthwith when the standard of living falls below a certain level but that the consequences are more long term: that is that the individual lives to age 55 instead of age 65 and so on. In the same way as with reduced food intake, the strain on the body occasioned by illnesses during the years of childhood and adolescence may lead to stunted growth and diminished average life-span. That even temporary undernourishment (because of harvest failure or epidemics) has consequences later in life for the capacity to work and for mortality means not only that average food intake during the growing years must be increased but also that its variations must be limited.

In any discussion of how poor countries can be made less poor, it is of the greatest importance to distinguish between countries where economic development has not yet started and those where development is under way. There is no recipe applicable to all countries. What is good for one country will not necessarily be good for all other countries. A policy that is good for getting development started will not necessarily be the right one for sustaining the process. We shall now pass on to a discussion of how development has proceeded in different parts of the world, in order to understand better what it is that starts and promotes development.

3. DEVELOPMENT PATTERNS AND DEMOGRAPHIC TRANSITIONS

In a majority of developing countries the demographic transition is well under way, which is demonstrated not only in a dramatic fall in mortality rates, but also in an ensuing long-term decline in fertility. However, in spite of this apparent world-wide fertility transition, there are, as shown by Geoffrey McNicoll in Chapter 8, great differences in the ways this transition is being achieved. McNicoll identifies four types of fertility transitions: traditional capitalist (which refers to Latin America), soft state (South Asia), radical devolution (China), and growth with equity (East Asia). In addition, a fifth pattern, lineage dominance, is identified for Sub-Saharan Africa, where there are only slight signs, if any, of a fertility transition.

In a broad sense, these differences in fertility patterns reflect differences in levels of economic performance and in types of economic organization. Although there is no mechanical link between economic performance and demographic transition, it is nevertheless possible to identify a certain correspondence between types of economic transition and types of fertility transition. The developing countries do not represent one coherent block of countries with identical socio-economic characteristics. Of course, they have never been identical, but it appears that today the world is becoming increasingly polarized and differentiated. At one extreme one finds fast-growing economies, which are in the midst of a full-blown socio-economic transformation, while at the other extreme there are societies which seem to be sinking deeper into economic decay. So, while there are indeed societies which are in a virtuous circle of growth, development, poverty alleviation, and fertility transition, there are also countries which seem to be trapped in a vicious circle of economic stagnation, social unrest, poverty, and even increasing fertility.

In schematic form, three types of socio-economic organization can be identified among the less developed countries: fast-growing dynamic, slow-growing stagnant, and non-growing chaotic types. The dynamic form connotes, in the first place, the prospering economies of East and South-East Asia that display a rate of economic growth never before experienced in world economic history (for a recent account, see World Bank 1993). These societies are being transformed into industrial societies with rapid urbanization and overall modernization in much the same way as Europe was after its Industrial Revolution, the major difference being the, now relatively much faster, economic transformation and demographic transition. Equally important, however, is the fact that average living standards have been substantially improved and that poverty has been dramatically reduced in a large number of countries. The demographic response to this transformation is unmistakable; in some countries the demographic transition is almost completed and in others the transition is well under way. In fact, it is hardly any longer accurate to speak of some of these countries as pre-industrial societies. In general terms, the fertility transition denoted by McNicoll as 'growth with equity' is one of economic betterment induced by a rise in agricultural productivity leading to a falling demand for children through changes in household structures or types of family organization.

The slow-growing stagnant type of economy is a semi-industrialized, largely agriculture-dependent society, albeit with a large urban sector. However, it is not a traditional, that is, pre-industrial society, since in large degree its present socio-economic structure is a product of a process of earlier economic growth. In parts the urban industrial sector is very modern and a growing number of people connected with this sector have a living standard quite similar to that of people in a comparable social position in developed countries. However, in contrast to the dynamic societies, there is also growing urban poverty, to which should be added a rural poverty of sizeable proportions. Although this society is characterized by a certain rate of growth as well as by a fertility transition, the link between the two is not always easy to identify.

The process of economic growth has led to widening income inequalities. Large sections of the population, mainly rural, have been left out of the modernization, so that the extent of poverty is increasing in spite of notable improvements in per-capita income. However, fertility has come down substantially since the 1960s and the fertility transition appears to be irreversible regardless of whether it happens under 'traditional-capitalist' (Latin American) or 'soft-state' (South Asian) forms. It is not easy, however, to determine whether the fertility transition has been propelled by the growth process, symbolized by the existence of a very modern urban industrial sector, in which the level of security is raised by an increased supply of regular job opportunities, or by changing forms of urban poverty. In either case, family organization is modified and the cost of children raised.

In the chaotic type of society modern economic growth is minimal or non-existent. Agricultural productivity is declining and the urban industrial sector is small and low-productive. There is no visible improvement in average living standards and the extent of extreme poverty appears to be rising. These societies are by no means traditional or stationary, but what they have in common is that, for various reasons, they have not entered the stage of 'modern economic growth', that is, the process of long-term economic growth and industrialization. In other words, they are basically pre-industrial. Although there are large and growing urban populations, the modern sector is small and functionally subordinated to other sectors. There are few signs of a substantial increase in regular job opportunities in a growing modern sector and jobs in the informal sector are casual and low paid. Thus, the extremely high level of insecurity persists and incentives to changes in family size remain weak.

This type of economy is particularly common in Sub-Saharan Africa, that is, the societies characterized by 'lineage dominance'. Since the 1950s there has hardly been any fall in fertility. Although it might be possible to forecast a fall in the coming decades as Africa is increasingly becoming land-scarce, leading to land privatization and a lessened demand for rural labour, it is impossible to know whether such a decline would follow 'traditional-capitalist' or 'growth-with-equity' paths, or whether Africa will follow a path of its own. It seems as if the economic opportunities opened up by the urban sector have neither reduced the benefits of larger families nor increased the costs of children. As Caroline Bledsoe's chapter indicates, the high level of insecurity in both urban and rural sectors appears to be a motive for keeping fertility rates high. Furthermore, the extended family persists in spite of urbanization. Questions of relevance here are whether this fertility pattern persists because of the type of family organization, and/or whether the family type persists because of the lack of sustained economic growth and industrialization. In other words, are there cultural and other obstacles to the diffusion of economic stimuli, or are there no economic stimuli?

So, in economic terms we find a type of country, mainly in East and South-East Asia, for which the connection between economic development and demographic transition is well established. At the other extreme we find a type, mainly in Africa, for which a persistence of high fertility rates corresponds with non-development. To some extent both types seem to confirm the view that development and economic betterment are the key to the demographic transition. But of course matters are much more complicated, which is shown not least by the fact that falling fertility may be associated also with increasing poverty and that urbanization can help to preserve 'traditional' family forms with high fertility. Fertility in parts of Latin America and South Asia seems to have fallen even among the poorer strata, presumably because of a dramatic change in family organization that follows from urbanization, which has raised the cost of children. It is not impossible, in fact, that part of the fertility transition even in South-East Asia can be attributed to similar mechanisms and the poverty growth associated with the closing of the land frontier.[2] In parts of Africa where the rural–urban link remains strong, urbanization seems to perpetuate a high level of fertility. Cash remittances from children who enter the urban sector can help reduce rural poverty.

The dramatic fertility transition in China is a particularly interesting case in point. Perhaps China can be treated as a special case, since obviously the fertility reduction has been instigated by enforcement of political decrees, that is, the 'radical-devolution' pattern depicted by McNicoll. One should be careful, however, not to overlook possible links to the apparent changes in family structure that followed from the rural transformation after 1955, under which the collective (village) and the commune (township) gradually took over responsibilities normally placed upon the family. The household ceased to exist as a form of organization for production and, thereby, the link was broken between production and reproduction. Obviously, the link between economic and demographic transitions will be put to a test now when the household has been re-established as a productive unit under the economic reform process. That the fertility decline is closely related to changing economic conditions is clearly indicated by the urban fertility pattern. The drastic increase in the female labour-force participation rate has greatly reduced the incentive to have large families. The important thing here is not to debate whether the fertility transition is due to family planning or to economic development, but to remember that both the fertility transition and the economic transformation were products of a command system, which had effects on both the timing and the strength of the fertility transition. The time-lag between economic changes and the fertility decline was reduced to a minimum and the strength of the fertility transition is unmatched even by the dynamic type of market economies in East Asia.

4. THE ROLE OF INSTITUTIONS

When analysing the connection between development and demographic responses it is imperative that those institutional factors are clarified which determine the links between economic and demographic behaviour. What we need to know is how different institutions influence economic behaviour and forms of organization so that the different types of behaviour and forms of organization specific to fertility can be identified. It is only then that we can understand how seemingly similar types of economic transformation can produce different fertility responses, and how seemingly similar fertility transitions can be associated with totally different types of economic transformation.

Institutions can be defined as 'rules, norms and customs, and their enforcement characteristics, which define the rights and obligations in human exchange'[3] or as 'clusters of behavioural rules governing human actions and relationships in recurrent situations' (McNicoll, Chapter 8). Since it is institutions, together with technology, that constrain all forms of human exchange, it is also by a change in institutions that the conditions of exchange can be altered. Institutional change is therefore the instrument of reform. But institutions can also be instrumental in obstructing reform. This means that, while it might be possible to identify key institutions needed to induce economic and demographic transitions, one will also find that there is an institutional reason why these key institutions are not implemented, or why similar reforms produce different outcomes. Furthermore, all institutions that influence the economic behaviour of individuals or households are not by definition economic institutions. Many forms of human interaction take place without any clear economic purpose. In pre-industrial societies, however, with limited division of labour, economic and other institutions are closely interlinked.

One way to identify the role of institutions is to say that they are established because there are transaction costs in exchange relations (Coase 1988). Transaction costs appear partly as information costs (imperfect or asymmetrical information) and partly as costs for carrying out exchange. They include the costs of establishing contact between economic actors (that is, gathering and diffusion of information) as well as costs for establishing contracts and costs for the control of the enforcement of contracts. High transaction costs mean that the level of insecurity in transactions is high. The more complex the exchange, the greater the transaction costs and the more likely it is for institutional arrangements to be established to reduce these costs, and hence reduce insecurity. If transaction costs are the source of institutions, it follows that institutions will have to become more complex and formalized as the complexity of the economy increases.

Institutions are devices that provide security in human interaction and which help in diffusing incentives. In a sense, institutions are always aimed at increasing security. But security has different connotations in different contexts. The reason why institutions differ is that security is a concept with different connotations between societies. In pre-industrial societies people live in close accord with nature and are highly dependent on variations in what nature gives. Without sufficient technological improvements people find themselves in a vicious circle of poverty, where the Malthusian trap closes time and again. Insecurity is extremely high and institutions that develop are in large degree risk-averting devices (see e.g. Scott 1976). Thus, institutions can be established as protective measures against a hostile environment. But institutions can also be established in order to encourage exchange and to make economic development possible by diffusion of incentives or by widening of the market.

The low level of technology and violent harvest fluctuations are major sources of insecurity in pre-industrial societies. Another factor may be found in the institutional structure. Obviously, without some institutional arrangement which reduces transaction costs by minimizing uncertainty people will be unwilling to enter into economic transactions. In societies with low levels of technology, in which the surplus-producing capacity is limited, institutions will be established to uphold short-term stability whereas institutions which encourage investments for future returns will be established only exceptionally. Few or no formal rules are needed when trade is arranged in local markets, and autonomous political and economic spheres hardly exist in such societies. Political authority is locally based and intertwined with and involved in the economic transactions.

Many institutions that look 'traditional' may obviously have modern origins. They can rather be seen as informal and local responses to more formal institutional arrangements that are established as societies change and the forms of exchange are being transformed. As exchange is widened, as markets begin to become integrated with other markets, it becomes necessary for property rights to be formalized. Property rights need to be defined for the simple reason that trade is by definition an exchange of property rights. When exchange is personal, property rights are understood and respected by force of custom. When exchange becomes impersonal it is important that the property rights of the parties involved are formally defined, enforced, and protected. Rule of law must replace personal rule. Thus, the state cannot be kept conceptually separated from the market. Every nationally integrated market requires at least a minimal state.

Two *essential* conditions must, by definition, be fulfilled here: the right to private property (inclusive of the right to use and exchange property and the right to generate income from the possession of property); and respect and protection of agreements and contracts. Thus, it is a fundamental duty of the state to establish and give credibility and legitimacy to laws, rules, and other institutional arrangements which guarantee property rights and which protect the validity of contracts. Only then will the level of security be high enough to allow more complex forms of exchange and long-term investment.

In the real world the problem is that such institutional devices seem to be established only exceptionally. The state has a particularly important role here. Although needed to guarantee the institutions that make exchange possible, the state also produces exchange-inhibiting institutions. As Douglass North (1979: 249) puts it: 'The creation of a state is an essential precondition for economic growth. The state, however, is the source of man-made economic decline.' This paradox is a paramount dilemma in development. On the one hand, development of poor countries is unfeasible without efficient economic institutions. On the other hand, it is a grim fact that 'predatory', 'praetorian, neo-patrimonial', 'prebendalist' and 'soft' states are more common than 'developmental' states. The development experience in many parts of the world obviously confirms a 'North's Law' that inefficiency is the rule, that is that the state, at least in pre-industrial societies, has an inherent tendency to reproduce institutions that impede economic growth. This also means that an institution that is efficient and serves a purpose at one time or place may be obsolete and development-inhibiting at another time or place. It also means that institutions are present at different levels. In every society there is a hierarchy of institutions, from the formal rules of the games that are set and enforced by the state to informal codes of conduct governing personal and local exchange. Thus, local institutions may be constrained by the formal institutions, but they may also have developed in response to, or for protection against, those formal institutions.

Thus, the informal or local institutions that persist and appear as 'traditional' institutions may very well be protective measures against predatory institutions imposed from above. Insecurity regarding property rights appears to be a development-inhibiting factor of fundamental importance. Ill-defined and badly protected property rights have undoubtedly made possible excessive taxation of agriculture, especially in Africa. The tax rates imposed on farmers have been far higher than can be justified as a 'fair' return. Africa's economic decline is primarily an effect of failures in agriculture. Inefficient institutions have brought about forms of organization and behaviour which aim at increasing short-term security. In contrast, these institutions do not encourage long-term investments in technologies which could help to break the vicious circle of low-level equilibrium of poverty.

When poverty persists, traditional patterns of organization persist and so do traditional forms of fertility behaviour. However, it is by no means clear that all 'traditional forms' are inherited from the past, since many of the insecurity-enhancing institutions are in fact modern creations. So what might seem to be expressions of a cultural heritage may just as well be a habitual way of adapting to new forms of insecurity. If this low-level equilibrium insecurity is eliminated, it is not at all certain that these traditional fertility patterns have any relevance whatsoever.

There is yet another aspect to property rights and security which needs to be emphazised here and that is the dispersion of property. It is nowadays well known that economic incentives are best diffused if there is an 'equality of opportunity' among potential economic actors. This means that people should enjoy equality in terms of opportunities of participating in economic life and to responding to economic incentives. Unless there is a fair amount of equality of opportunity, there is no reason to expect that economic reforms will be received equally among economic actors.

The question of equal opportunities is closely related to the issue of access to property, of the possibility of deriving an income from property. Insecurity and failure to respond to economic incentives is often associated with an inadequate property base. Landlessness, various forms of tenancy, and fragmentation of smallholdings are all factors which tend to keep insecurity levels high. Other factors are insufficient access to credit, markets, technology, and education. Thus, insecurity may be associated with insecure formal property rights, but also with an absolute property shortage. In many developing countries institutional structures are protected by the state which deprive large sections of the population of any chance of participating in a process of sustained economic development. Consequently, they can hardly be expected to participate freely in a demographic transition.

As has been shown in the works of Adelman *et al.* (1992), Oshima (1992), and others, access to property is of fundamental importance in economic development. An aspect which is seldom mentioned as a 'lesson from East Asia' is that the thriving market economy has been based on the principle of dispersion of property rights. It is often ignored that a market economy requires 'equality of opportunity'. In order to create equal opportunities, some form of redistribution of economic resources is usually needed. In Taiwan, income distribution was made relatively equal, mainly as a result of the implementation of a comprehensive land reform. Thus, we see that redistribution of the land was an important factor in creating equal opportunities. In most other underdeveloped countries inequality of opportunity has been extremely high right from the start. In these cases, industrialization has served to reinforce inequality and social cleavages more than anything else. As a result, governments have gradually been losing legitimacy, sometimes to the point where the entire political sphere is being perverted by private interests. All this leads to the conclusion that institutional arrangements aimed at redistribution may be needed for growth of the market economy. In part, the plea for redistribution is an argument for land reform. But land reform is not the key to development in all countries, nor can land reforms alone solve the insecurity problem. It seems that in Africa insecure property rights are more important than the unequal distribution of income. It is also important that land reforms should be accompanied by other institutional arrangements. Redistribution of land rights without simultaneous efforts with respect to credit, marketing, and technology would not do much good; it might only equalize poverty and maintain low levels of security. What is important here is not only redistribution, the establishment of property rights, and the elimination of discriminatory pricing against agriculture, but also an emphasis on the development of institutions which help in diffusing agricultural technology.

The link with McNicoll's types of fertility transition is clear enough. We are now able to see which institutional arrangements are contained in the process of 'growth with equity'. Thus, it is not economic development in broad or vague terms which has fostered the fertility transition, but a specific form of economic development, with specific institutional arrangements, which has induced changes in agricultural productivity, household structure, and the demand for children.

What, then, can we learn from the ways in which the insecurity problem has been handled in history? Is there a lesson about the relationship between economic transformation and demographic transition to be drawn from success stories in past and present times? Basically, one can say that the industrial transformation in Western Europe originated in the improvements in agricultural productivity. Institutional reforms which defined, protected, and diffused property rights made possible investments in technologies that brought an increase in per-capita production. Agricultural surpluses were no longer consumed by growing populations and the element of extreme insecurity was dramatically reduced. Productivity improvements in agriculture also facilitated a process of industrialization through transfers of capital and redundant labour to the industrial sector. Increasing rural incomes also provided a market for industrial goods. Shifts in the demand for agricultural labour and the new opportunities opened up by the urban sector induced the fertility transition.

Thus, it was not until the level of fundamental security was raised that the process of economic development and fertility transition actually started. The transition in East Asia, in fact, followed the same pattern. The security level was raised by institutional reforms that made possible a revolution in agricultural productivity. The East Asian 'growth-with-equity' model has in common with Europe the fact that its agricultural revolution was founded on institutions that defined, protected, and diffused property rights.

The European and East Asian models indicate that there are key institutions which create the connection between economic and demographic transitions. The Chinese fertility transition, however, indicates that substitution is possible. Fertility can be brought down by other institutional measures, notably by means of coercion. However, the coercive element has two dimensions: one, birth control, which affects fertility directly; and another, collectivization, that has an indirect effect on fertility by creating new forms of economic organization. So also in China the economic and demographic transitions go together; in fact, they also go together in the precise sense that they are simultaneous. The apparent success of the Chinese substitution model should not be interpreted as an argument in favour of coercion. As Sen shows in Chapter 3, fertility has been reduced substantially in the poor Indian state of Kerala too, and this fertility reduction has been achieved without using coercion. The case of Kerala would indicate that conditions had been created which encouraged 'reasoned decisions and enlightened rational behaviour'. Such conditions could only have been established if the basic security problem had been solved. In fact, it is not unlikely that in China too it became possible to take such reasoned decisions about fertility behaviour after the basic security problem had been solved.

What population institutions are required to make the transition as fast and as smooth as possible? Does decline in fertility take place earlier in societies with high marriage ages, a high proportion of unmarried people, and nuclear families as the predominant form of family given similar economic development? In Western Europe, where family size decreased in all countries except Portugal and Ireland prior to 1910, the marriage age prior to the fertility transition was 23–8 years for women, 10–25 per cent of women never married, and the nuclear family was predominant. In Eastern Europe, where the fertility transition began after 1910, the pre-transition marriage age was 19–22 years, 2–5 per cent of women never married, and the extended family was a common form of family. In Asia and Africa, finally, where the large family has been predominant, where the marriage age as a rule has been under 18 years, and where fewer than 1 per cent of women have remained unmarried, the fertility transition occurred mainly after 1950, if it has begun at all. In each one of these regions, also, the same pattern can be discerned (Coale 1992). The fertility transition started where the marriage age was at its highest and where the proportion of unmarried people was highest—in other words, in the societies where the incentive to limit population growth was strongest. For those societies dominated prior to the transition by extended families, the trend has been towards nuclear families. Is the nuclear family a key institution for decreased family size? Must the relations between men and women be brought into the picture as well? Can social structures at village and community level influence the fertility transition and possibly substitute for 'disadvantageous' family forms? What significance do national political and legal systems, and so on, possess?

In order to elucidate these questions we must analyse the important functions of different family systems. These differ with regard to the way in which transfers take place between the generations, how new households are established, the role difference between men and women, and the marital fertility responsiveness to the changing economics of children. In Western Europe, population growth was 'controlled' by variation of the marriage age and of the proportion of people unmarried. The nuclear family was predominant and marital fertility was fairly unaffected by changes in living standards during the pre-industrial period. Mortality varied in the long term, but fairly independently of living standards. There was possibly a connection between living standards and mortality, but with a long time-lag. The family was a unit of both consumption and production. Family agriculture was predominant and the tasks performed by men and women were different. Farmers strove to become freeholders so that they could hand over the farm to an adult son in exchange for his parents being looked after in old age. The overlap between generations was low because of the high marriage age and the low average life-span. Insecurity was mainly the product of dependence on the harvest outturn. Storage was difficult and sources of alternative income few. To even out consumption as much as possible, agriculture was organized on strip-cultivation lines. In other words, what we are looking at is a society which was endeavouring to solve the problems of uncertainty on the production side. When this was done, 'modern economic growth' took place and the family structure changed, starting in areas with nuclear families and high age at marriage, in areas with secure property rights, and so on. Also, in the parts of Asia where developments have been most rapid, the economic and social structure has been quite similar. In other parts of the world which could be characterized as high-risk or chaotic societies, in particular in Africa, the security problem has been solved mainly on the social side by large families, large networks with links both to the rural and the modern sector of the economy. Will family forms and access to private property in African countries be transformed as in China? Will all countries converge into the nuclear–family system? The answer is that it may very well be possible to substitute for these institutions by arrangements at local level, as McNicoll clearly demonstrates.

5. WHAT WE HAVE LEARNT

At the UN conference in Bucharest in 1974 the conclusion, in somewhat simplified terms, was that socio-economic development is the best contraceptive. In Mexico in 1984 it was that contraceptives are the best contraceptive—that is, that family planning is the main key to lower fertility. Since then this conclusion has been called into question repeatedly (see e.g. Hill 1992). What will be the conclusion of the forthcoming UN conference in Cairo in 1994? Will it be that education is the best contraceptive? Will it be that the role of women must be strengthened? Will a single factor be presented as the main key in limiting population increase again? A major result of the seminars and discussions preceding this book is that there is no single solution that could be applied to all poor countries. Any successful policy must take into account the different characteristics of each country in terms of its stage of development, its culture, and its social and economic institutions. There are several roads to development. However, whatever the route and whatever institutional reforms that are implemented, the development and the demographic transition start with security in the short run, that is, from year to year. Thus, one might say that security is the best contraceptive.

The lessons for Latin America, South Asia and Africa ought to be that, before one can expect any lasting effect from birth-control programmes or any other population programmes, the preconditions for economic development ought to be altered. In all of these cases development is obstructed (Africa) or distorted (Latin America, South Asia) by the lack of rural institutional changes which can create preconditions for a long-lasting improvement in agricultural productivity and thereby industrialization and modern forms of economic development. Economic development can take many divergent forms and so can the patterns of fertility transition. Nor can development models be transferred easily between countries. However, historical experience does show that, before one can expect either long-term economic growth or a demographic transition, the basic insecurity problem in the pre-industrial society has to be resolved. The European process of industrialization started with security-enhancing technological and institutional improvements in agriculture, and so did the the 'miracle' of East Asia.

To ensure security and well-being from year to year, it is imperative that the causes of rural poverty should be attacked vigorously and that priority should be given to agriculture in national and local development strategies. Particular efforts must be made to promote and support the viability of small-scale agriculture through improvements in credit institutions, price incentives, information, and education services. Institutional reforms to redistribute and ensure rights to land are necessary. Typically, both to create short-term security and long-term development, there is a need for a mixture of policies combining incentives and institutional reforms in agriculture with improved employment opportunities for women, improved educational programmes, a more extensive availability of basic-need goods such as potable water and household fuel, and credit, savings, and insurance facilities. This is the true basis for both development and the fertility transition.

In order to create development policies and institutional reforms for individual countries, education and research programmes in the poorest parts of the world must be improved. To create such programmes, and to avoid centralized development plans that do not take into account the different constraints and possibilities of each country, resources from centralized international organizations must be decentralized to the developing part of the world.

NOTES

1. The influence of economic variations on fertility, mortality, and nuptiality in pre-industrial Europe and other parts of the world has been shown in a large number of studies by R. D. Lee, P. Galloway, T. P. Schultz, D. Weir, T. Bengtsson, and others (see Lee 1993). For the decline in the effects of economic variations on mortality in different ages during the transitional stage, see Bengtsson and Ohlsson (1985).

2. For a discussion about Thailand, see Hetaserani and Roumasset (1991).

3. This definition corresponds roughly to the definition used by Douglass North (see North 1990).

REFERENCES

Adelman, I., Taft Morris, C., Fetine, H., and Golan-Hardy, E. (1992), 'Institutional Change, Economic Development, and the Environment', *Ambio*, 21.

Bengtsson, T. (1992), 'Lessons from the Past: The Demographic Transition Revised', *Ambio*, 21.

Bengtsson, T., and Ohlsson, R. (1985), 'Age-Specific Mortality and Short-Term Changes in Standard of Living: Sweden, 1751–1959', *European Journal of Population*, 1: 309-26

Coale, A. (1992), 'Age of Entry into Marriage and the Date of the Initiation of Voluntary Birth Control', *Demography*, 29.

Coase, R.H. (1988), *The Firm., the Market and the Law* (Chicago).

Dasgupta, P. (1992), 'Population, Resources and Poverty', *Ambio*, 21.

Fridlizius, G. (1984), 'The Mortality Decline in the First Phase of the Demographic Transition: Swedish Experiences', in T. Bengtsson, G. Fridlizius, and R. Ohlsson (eds.), *Pre-Industrial Population Change* (Lund).

Hetaserani, S., and Roumasset, J. (1991), 'Institutional Change and the Demographic Transition in Rural Thailand', *Economic Development and Cultural Change*, 39.

Hill, K. (1992), 'Fertility and Mortality Trends in the Developing World', *Ambio*, 21.

Lee, R. D. (1993), 'Inverse Projections and Demographic Fluctuations: A Critical Assessment of New Methods', in D. S. Reher, and R. Schofield (eds.), *Old and New Methods in Historical Demography* (Oxford).

North, D. C. (1979), 'A Framework for Analysing the State', *Explorations in Economic History*, 16:349.

North, D. C. (1990), *Institutions, Institutional Change and Economic Performance*, (Cambridge).

Oshima, H. (1992), 'Impact of Economic Development on Labor Markets, Education and Population in Asia', *Ambio*, 21.

Perrenoud, A. (1984), 'The Mortality Decline in a Long-Term Perspective', in T. Bengtsson, G. Fridlizius, and R. Ohlsson (eds.), *Pre-Industrial Population Change* (Lund).

Schofield, R. (1984), 'Population Growth in the Century after 1750: The Role of Mortality Decline', in T. Bengtsson, G. Fridlizius, and R. Ohlsson (eds.), *Pre-Industrial Population* Change (Lund).

Scott, J. (1976), *The Moral Economy of the Peasant* (New Haven, Conn., and London).

Wrigley, E. A., and Shofield, R. S. (1981), *The Population History of England 1541–1871. A Reconstruction* (London)

World Bank (1993), *The East Asian Miracle* (Oxford).

2

The Environmental Resource Base and Human Welfare

Partha Dasgupta[*], Carl Folke[†], and Karl-Göran Mäler[‡]

1. ECOSYSTEMS: FUNCTIONS AND SERVICES

The ecological services we rely upon are produced by *ecological systems* (or *ecosystems* for short). These services are generated by interactions among organisms, populations of organisms, communities of populations, and the physical and chemical environment in which they reside. Ecosystems are involved in a number of functions. They also offer a wide range of services. Many are indispensable, as they provide the underpinning for all human activities. So they are of fundamental value. Among other things, ecosystems are the sources of water, of animal and plant food, and of other renewable resources. They also maintain a genetic library, sustain the processes that preserve and regenerate soil, recycle nutrients, control floods, filter pollutants, assimilate waste, pollinate crops, operate the hydrological cycle, and maintain the gaseous composition of the atmosphere (see e.g. Odum 1975; Folke 1991; de Groot 1992; Ehrlich and Ehrlich 1992; and Section 2 below). The totality of all the ecosystems of the world represents a large part of what we may call our *natural capital base*.[1] For vividness, we will often refer to it in what follows as the *environmental resource base*.

Since the services this base provides for us are essential for our survival, it would clearly be prudent to monitor it in much the same way that we routinely monitor our manufactured capital stocks, such as roads, buildings, and machinery. Unhappily, this has not been standard practice.

[*] Professor of Economics, University of Cambridge.
[†] Associate Professor of Systems Ecology, Stockholm University.
[‡] Professor of Economics, Stockholm School of Economics.

Even today it is not conducted in any systematic way. Instead, reliance is often placed on time trends in an economy's gross outputs (for example, agricultural crops, fisheries, forest products) and in their prices for obtaining a sense of whether the resource base is becoming depleted. This is a mistake. Agricultural output, for example, could in principle display a rising trend even while the soils are being mined. This point is important to bear in mind. The environmental resource base is a dynamic and complex living system, consisting as it does of biological communities that interact with the physical and chemical environment in time and space. Moreover, the interactions are often non-linear (see Section 2 below). Therefore, the state of a resource base can display threshold effects. This in turn means that there can be discontinuities lying in wait in the flow of services that we enjoy.

Degradation of the environmental resource base (for example, excessive resource extraction, severe land use, and so forth) not only affects the quantity and quality of the services that are produced by ecosystems; it also challenges their resilience. Resilience is the capacity of a system to recover from perturbations, shocks, and surprises. An ecosystem's resilience is its capacity to absorb disturbances without undergoing fundamental changes. If a system loses its resilience, it can flip to a wholly new state when subjected to even a small perturbation. Thus, the economist's panacea, the view that there are unending substitution possibilities among various resources, that society will be able to move smoothly from one resource base to another as each is degraded beyond usefulness, is at odds with fundamental ecological truths. That ecosystems have limited resilience lies at the heart of the matter. The problem is that reductions in resilience are not easily observable. This is what makes the subject of ecological economics so very difficult.

The *carrying capacity* of an ecosystem is the maximum stress it is capable of absorbing without it flipping to a vastly different state (see Section 3 below). Ecosystems are endemically subject to natural shocks and surprises (for example, fires, storms, and so on). This means that it would be incorrect to regard them as fixed stocks of capital that can be relied upon to provide us with a steady flow of resources. Our natural capital base has evolved over millions of years. It has adapted to modifications and fluctuations in the background environment. The self-organizing ability of ecosystems determines its capacity to respond to the perturbations they are continually subjected to (see e.g. Holling *et al.* 1994).

Biological diversity (or biodiversity for short) plays two central roles in the evolution of ecosystems. First, it provides the units through which both energy and materials flow, thus giving the system its functional properties. Secondly, it provides the system with resilience (Wilson 1992; Solbrig 1993).

It is as well to emphasize that an ecosystem's carrying capacity does not remain constant, but is subject to change, usually in ways that are unpredictable. This is because ecosystems evolve continually. Economic policies that apply fixed rules so as to achieve constant yields (for example, fixed sizes of cattle herds or wildlife, or fixed sustainable yield of fish or wood) can be expected to lead to a reduction in an ecosystem's resilience. Large reductions in resilience would mean that the system could break down in the face of disturbances that earlier would have been absorbed (Holling *et al.* 1994).

Grazing in the semi-arid grasslands of East and South Africa offers a good illustration of these observations. Under natural conditions these grasslands are periodically subject to intense grazing by large herbivores. The episodes are much like pulsations, and they lead to a dynamic balance between two functionally different groups of grasses. One group, which tolerates grazing and drought, has the capacity to hold soil and water. The other is productive in terms of plant biomass and enjoys a competitive advantage over the first group during those periods when grazing is not intensive. In this way, a diversity of grass species is maintained. This diversity serves two ecological functions: productivity and drought protection. Grazing by large herbivores that periodically shift from intense pulses to durations when recovery is permitted forms a part of the overall dynamics of the ecosystem. However, when fixed management rules are applied there (for example, the stocking of ranch cattle at a sustained and moderate level), it can mean a shift from the periodically intense pulses of grazing to a more modest, but persistent, level of grazing. The latter mode, occasioned by deliberate economic policy, supports the competitive advantage of the productive, but drought-sensitive, grasses at the expense of the drought-resistant grasses. This, in turn, means that the functional diversity we spoke of earlier is reduced, and the grassland can flip and come to be dominated by woody shrubs that are of low value for grazing.

There are many examples of management policies that share this feature. They occur in agriculture, fisheries, range land, and forest management (e.g. Walters 1986; Regier and Baskerville 1986; Trenbath, Conway, and Craig 1990; Folke and Kautsky 1992; Holling *et al.* 1994; Perrings and Walker, 1994). The significance and value of the 'infrastructural composition' of the environment and its dynamics have received little attention in economics. As we noted earlier, this is surprising, since the production capacity of the resource base constitutes the foundation for human life and its development. A challenge emanating from this is to estimate the stream of benefits that are forgone when there is a contraction in an ecosystem's resilience. The major difficulty in any such exercise lies in the limitations of the commonly used models in applied environmental economics. Typically, they do not entertain non-linearities in the ecological processes that are modelled. Typically also, these models insulate the economic system from its environment and ignore the evolutionary tendencies of the resource base, thereby abstracting from some of the most important characteristics of self-organizing systems (that is, the existence of environmental feedbacks, thresholds, and discontinuities). Because of both the evolutionary nature of an ecosystem's responses to changes in the background environment (including human encroachment) and the existence of threshold effects, these feedbacks are largely unpredictable, except within ranges in which the system exhibits local stability (Perrings *et al.* 1994). In other words, certain dynamic effects tend to be ignored (and often, not even perceived) when economists attempt to aggregate values derived from partial observations of expenditure patterns in the wake of some change in the level of ecological resources or services. In the remainder of this overview, we will try to identify the implications of these observations for economic policy in poor countries.

2. THE RESOURCE BASE AND HUMAN WELFARE

People in poor countries are for the most part agrarian and pastoral folk. In 1988 rural people accounted for about 65 per cent of the population of what the World Bank classifies as low-income countries. The proportion of total labour force in agriculture was a bit in excess of this. The share of agriculture in gross domestic product in these countries was 30 per cent. These figures should be contrasted with those from industrial market economies, which are 6 per cent and 2 per cent, respectively.[2]

Poor countries are for the most part biomass-based subsistence economies, in that their rural folk eke out a living from products obtained directly from plants and animals. For example, in their informative study of life in a micro-watershed of the Alaknanda River in the central Himalayas in India, the (Indian) Centre for Science and Environment (CSE 1990) reports that, of the total number of hours worked by the villagers sampled, 30 per cent was devoted to cultivation, 20 per cent to fodder collection, and about 25 per cent was spread evenly between fuel collection, animal care, and grazing. Some 20 per cent of time was spent on household chores, of which cooking took up the greatest portion, and the remaining 5 per cent was involved in other activities, such as marketing. In their work on Central and West Africa, Falconer and Arnold (1989) and Falconer (1990) have shown how vital are forest products to the lives of rural folk. Come what may, poor countries can be expected to remain largely rural economies for a long while yet.

The dependence of poor countries on their natural resources, such as soil and its cover, water, forests, animals, and fisheries, should be self-evident: ignore the environmental resource base, and we are bound to obtain a misleading picture of productive activity in rural communities there. Nevertheless, if there has been a single thread running through forty years of investigation into the poverty of poor countries, it has been the neglect of this resource base. Until very recently, environmental resources made but perfunctory appearances in government planning models, and they were cheerfully ignored in most of what goes by the name of development economics.[3]

The situation is now different, and today no account of economic development would be regarded as adequate if the environmental resource base were absent from it. This brief chapter, therefore, raises some aspects of the environment and emerging development issues.

Environmental problems are almost always associated with resources that are regenerative (we could call them renewable natural resources), but that are in danger of exhaustion from excessive use.[4] The earth's atmosphere is a paradigm of such resources. In the normal course of events, the atmosphere's composition regenerates itself. But the speed of regeneration depends upon, among other things, the current state of the atmosphere and the rate at which pollutants are deposited. It also depends upon the nature of the pollutants. (Smoke discharge is different from the release of chemicals or radioactive material.) Before all else, we need a way of measuring such resources. In the foregoing example we have to think of an atmospheric quality index. The net rate of regeneration of the stock is the rate at which this quality index changes over time. Regeneration rates of atmospheric quality are complex, often ill-understood matters. This is because there is a great deal of synergism associated with the interaction of different types of pollutants in the atmospheric sink, so that, for example, the underlying relationships are almost certainly non-linear and, for certain compositions, perhaps greatly so. What are referred to as 'non-linear dose-response relationships' in the ecological literature are instances of this.[5] But these are merely qualifications, and the analytical point we are making, that pollution problems involve the degradation of the environmental resource base, is both true and useful (see Ehrlich, Ehrlich, and Holdren 1977).

Animal, bird, plant, and fish populations are other examples of renewable natural resources, and there are now a number of studies addressing the reproductive behaviour of different species under a variety of 'environmental conditions, including the presence of parasitic and symbiotic neighbours.[6] Land is also such a commodity, for the quality of arable and grazing land can be maintained only by careful use. Population pressures can result in an extended period of over use. By over use we mean not only an unsustainable shortening of fallow periods, but also deforestation, and the cultivation and grazing of marginal lands. This causes the quality of land to deteriorate, until it eventually becomes a wasteland.

The symbiotic relationship between soil quality and vegetation cover is central to the innumerable problems facing Sub-Saharan Africa, most especially the Sahel.[7] The management of the drylands in general has to be sensitive to such relationships. It is, for example, useful to distinguish between, on the one hand, a reduction in soil nutrients and humus, and, on the other, the loss of soil due to wind and water run-off. The depletion of soil nutrients can be countered by fertilizers (which, however, can have adverse effects elsewhere in the ecological system), but in the drylands a loss in topsoil cannot be made good. (In river valleys the alluvial topsoil is augmented annually by silt brought by the rivers from mountain slopes. This is the obverse of water run-off caused by a lack of vegetation cover.) Under natural conditions of vegetation cover it can take anything between 100 and 500 years for the formation of 1 cm of topsoil. Admittedly, what we are calling 'erosion' is a redistribution of soil. But even when the relocation is from one agricultural field to another, there are adjustment costs. Moreover, the relocation is often into the oceans and non-agricultural land. This amounts to erosion.[8] Soil degradation can also occur if the wrong crops are cultivated. Contrary to general belief, in subtropical conditions most export crops tend to be less damaging to soils than are cereals and root crops. (Groundnuts and cotton are exceptions.) Many export crops, such as coffee, cocoa, oil palm, and tea, grow on trees and bushes that enjoy a continuous root structure and provide continuous canopy cover. With grasses planted underneath, the rate of soil erosion that is associated with such crops is known to be substantially less than the rate of erosion associated with basic food crops (see Repetto 1988: table 2). But problems are piled upon problems in poor countries. In many cultures the men control cash income while the women control food. Studies in Nigeria, Kenya, India, and Nepal suggest that, to the extent that women's incomes decline as the proportion of cash-cropping increases, the family's nutritional status (most especially the nutritional status of children) deteriorates (Gross and Underwood 1971; von Braun and Kennedy 1986; Kennedy and Oniang'o 1990). The indirect effects of public policy assume a bewildering variety in poor countries, where ecological and technological factors intermingle with norms of behaviour that respond only very slowly to changing circumstances.[9]

The link between irrigation and the process by which land becomes increasingly saline has also been much noted in the ecological literature (see Ehrlich, Ehrlich, and Holdren 1977). In the absence of adequate drainage, continued irrigation slowly but remorselessly destroys agricultural land owing to the salts left behind by evaporating water. The surface area of agricultural land removed from cultivation world-wide through salinization is thought by some to equal the amount added by irrigation (see United Nations 1990). Desalinization of agricultural land is even today an enormously expensive operation.

The environment is affected by the fact that the rural poor are particularly constrained in their access to credit, insurance, and capital markets. Because of such constraints, domestic animals assume a singularly important role as an asset (see e.g. Binswanger and Rosenzweig 1986; Rosenzweig and Wolpin 1985; Hoff and Stiglitz 1990; Dasgupta 1993). But they are prone to dying when rainfall is scarce. In Sub-Saharan Africa farmers and nomads, therefore, carry extra cattle as an insurance against droughts. Herds are larger than they would be were capital and insurance markets open to the rural poor. This imposes an additional strain on grazing lands, most especially during periods of drought. That this link between capital and credit markets (or rather, their absence) and the degradation of the environmental resource base is quantitatively significant (World Bank 1992) should come as no surprise. The environment is itself a gigantic capital asset. The portfolio of assets that a household manages depends on what is available to it. In fact, one can go beyond these rather obvious links and argue that even the fertility rate is related to the extent of the local environmental resource base, such as fuel-wood and water sources. Later in this chapter (Section 5) we will see not only why we should expect this to be so, but we will also study its implications for public policy (see also Dasgupta 1993).

Underground basins of water have the characteristic of a renewable natural resource if they are recharged over the annual cycle. The required analysis is a bit more problematic though, in that we are interested in both its quality and its quantity. Under normal circumstances, an aquifer undergoes a self-cleansing process as pollutants are deposited into it. (Here the symbiotic role of microbes, as in the case of soil, is important.) But the effectiveness of the process depends on the nature of pollutants and the rate at which they are discharged. Moreover, the recharge rate depends not only on annual precipitation and the extent of underground flows, but also on the rate of evaporation. This in turn is a function of the extent of soil cover. In the drylands, reduced soil cover beyond a point lowers both soil moisture and the rate of recharge of underground basins, which in turn reduces the soil cover still more, which in turn implies a reduced rate of recharge, and so on.[10] With a lowered underground water table, the cost of water extraction rises.

In fact, aquifers display another characteristic. On occasion the issue is not one of depositing pollutants into them. If, as a consequence of excessive extraction, the ground-water level is allowed to drop too low, there can be salt-water intrusion in coastal aquifers, and this can result in the destruction of the basin.

Since environmental resources, such as forests, the atmosphere, and the seas, are multifunctional, they have multiple competing uses. This accentuates management problems. Thus, forests are a source of timber, bark, saps, and, more particularly, pharmaceuticals. Tropical forests also provide a habitat for a rich genetic pool. In addition, forests influence local and regional climate, preserve soil cover on site, and, in the case of watersheds, protect soil downstream from floods. Increased run-off of rainwater arising from deforestation helps strip soil away, depriving agriculture of nutrients and clogging water reservoirs and irrigation systems. The social value of a forest typically exceeds the value of its direct products, and on occasion greatly exceeds it (see Ehrlich, Ehrlich, and Holdren 1977; Dasgupta 1982; Hamilton and King 1983; Anderson 1987).

It is as well to remember that the kinds of resources we are thinking of here are on occasion of direct use in consumption (as with fisheries), in production (as with plankton, which serves as food for fish species), and sometimes in both (as with drinking and irrigation water). Their stock is measured in different ways, depending on the resource: in mass units (for example, biomass units for forests, cow dung, and crop residues), in quality indices (for example, water and air quality indices), in volume units (e.g. acre-feet for aquifers), and so on. When we express concern about environmental matters, we in effect point to a decline in their stock. But a decline in their stock, on its own, is not a reason for concern. This is seen most clearly in the context of exhaustible resources, such as fossil fuels. Not to reduce their stocks is not to use them at all, and this is unlikely to be the right thing to do. Much has been written about the basis upon which their optimal patterns of use should be discussed (see e.g. Dasgupta and Mäler 1993). But even a casual reading of the foregoing examples suggests that a number of issues in environmental economics are 'capital-theoretic'. These issues will be the substance of Section 7 of this chapter.[11]

3. NEEDS, STRESS, AND CARRYING CAPACITY: LAND AND WATER

How much land does a man need? Here we will provide some orders of magnitude in tropical subsistence economies, and for simplicity we will ignore uncertainty in production.[12]

Rice cultivation in the drylands using conventional techniques requires something like 130 person-days of labour time per hectare each year, and it yields about 15 billion joules (GJ), or 1.000 kg of rice. If the average energy input in cultivation per working day is taken to be 3 million joules (MJ), total energy required for cultivation amounts to 390 MJ, or 0.39 GJ, per hectare over the year. Therefore, the net energy produced amounts to 14.61 GJ. Assuming an individual's energy requirements is 2,200 kcal per day, a family with five members would require 17 GJ of food energy per year.[13] This in turn means that the family would need approximately 1.2 ha of land to remain in energy balance. Inversely, a family of five would be the *carrying capacity* of 1.2 ha. Total work input on this amount of land is 0.39*1.2 GJ, or approximately 0.5 GJ. Therefore, the ratio of energy output to energy input in this form of cultivation is 34:1. This is quite high, and compares very favourably with the energy output–input ratios associated with the technologies available to hunter-gatherers, pastoralists, and food-garden systems in fertile tropical coastal areas.[14]

Crude, but revealing, calculations of this kind can also be done for water requirements. While 70 per cent of the earth's surface is covered by water, about 98 per cent of this is salt-water. Most of the earth's fresh water is stored as polar ice-caps and in underground reservoirs. (Only about 0.015 per cent is available in rivers, lakes, and streams.) It is distributed most unevenly across regions.

The three sources of water for any given territory are transfrontier aquifers, rivers from upstream locations, and rainfall. The water that is available from precipitation comes in two forms: soil moisture, and the annual recharge of terrestrial water systems (aquifers, ponds, lakes, and rivers). Rain-fed agriculture consumes an amount of water roughly proportional to the produced biomass (the water is returned to the atmosphere as plant evapotranspiration). Water can be recycled, and so the water utilization rate can be in excess of the water supply. (Israel, Libya, and Malta have utilization rates well in excess of annual water supplies.) The problem is that, in semi-arid and arid regions, losses due to evaporation from natural vegetation and wet surfaces are substantial, and not much effort is made in the poor drylands to develop technologies for reducing them (for example, improved designs of tanks and reservoirs). The mean annual precipitation divided by mean annual potential evapotranspiration is less than 0.03 in hyper-arid regions (annual rainfall less than 10 cm); between 0.03 and 0.20 in arid regions (annual rainfall between 10 and 30 cm); between 0.20 and 0.50 in semi-arid regions (annual rainfall between 20 and 50 cm); and between 0.50 and 0.75 in sub-humid regions (annual rainfall between 50 and 80 cm). According to most classifications, this set of regions comprises the drylands. Within the drylands, rain-fed agriculture is suited only to sub-humid regions. Occupying about a third of the earth's land surface, the drylands are the home of some 850 million people (see Dixon, James, and Sherman 1989: 3).

Losses due to evapotranspiration are dependent upon soil cover. It would appear to be at a maximum when the soil moisture is at full capacity (the so-called 'field capacity') and the soil is fully covered with vegetation (see Penman 1956). So a reduction in cover would lower evapotranspiration. But this would be so only up to a point: ecosystems are structurally stable only within limited regions of the space of their underlying parameters.[15] The idea of 'thresholds effects' is an instance of this. Thus, beyond a point, losses due to evaporation in the drylands are accelerated by disappearing biomass. For example, only about 10–20 per cent of rainfall finds use in the production of vegetation in the Sahelian rangelands (where the annual rainfall is in the range 10–60 cm). Some 60 per cent is returned to the atmosphere as unproductive evaporation. Irrigation schemes in the drylands, bringing water from distant parts, is unlikely to be cost-effective. This is a solution more appropriate to temperate zones. It has been argued that the proportion of rainfall in the drylands that is productive can be increased to 50 per cent if vegetation is allowed to grow, and if suitable catchments are constructed (see Falkenmark 1986; Barghouti and Lallement 1988).

Something like 1,250 cubic metres of water per person is required annually for the supply of habitats and for the production of subsistence crops in the drylands. This does not include the water that is required for municipal supplies, for industry, and for the production of cash crops. (Agriculture currently uses about 75 per cent of the world's use of fresh water, industry about 20 per cent, and domestic activities the remaining 5 per cent.) A community experiences water stress if, for every 1 million cubic metres of water available annually for use, there are 600–1,000 people having to share it. When more than 1,000 persons are forced to share every 1 million cubic metres of water annually, the problem is one of severe shortage. Currently, well over 200 million people in Africa are suffering from water stress or worse (see Falkenmark 1989). The tangled web of population growth, deforestation, water stress, and land degradation defines a good deal of the phenomenon of destitution in the world as we now know it.

4. INSTITUTIONAL FAILURE AND POVERTY AS CAUSES OF ENVIRONMENTAL DEGRADATION

Understanding the ultimate causes behind this degradation requires the inclusion of environmental resources in economic analysis. The main causes are institutional failure and poverty.

The early literature on the subject identified failure of market institutions as the underlying cause of environmental problems (e.g. Pigou 1920; Meade 1973; Mäler 1974; Baumol and Oates 1975; Dasgupta and Heal 1979). Indeed, more often than not environmental economics is even today regarded as a branch of the economics of externalities. Recently, however, certain patterns of environmental deterioration have been traced to inappropriate government policies, not market failure (e.g. Feder 1977; Dasgupta 1982; Mahar 1988; Repetto 1988; Binswanger 1989; Dasgupta and Mäler 1991). Taken together, they reflect institutional failures. They will be the object of study in Section 5 of this chapter, where we will place these matters within the context of the thesis that environmental degradation is a cause of accentuated poverty among the rural poor in poor countries.

At the same time, poverty itself can be a cause of environmental degradation. This reverse causality stems from the fact that, for poor people in poor countries, a number of environmental resources are complementary in production and consumption to other goods and services, while a number of environmental resources supplement income, most especially in times of acute economic stress (see e.g. Falconer and Arnold 1989; Falconer 1990). This can be a source of cumulative causation, where poverty, high fertility rates, and environmental degradation feed upon one another. In fact, an erosion of the environmental resource base can make certain categories of people destitute even while the economy on average grows (see Dasgupta 1993: ch. 16).

These two causes of environmental degradation (namely, institutional failure and poverty) pull in different directions, and are together not unrelated to an intellectual tension between concerns about externalities (such as, for example, the increased greenhouse effect, acid rain, and the fear that the mix of resources and manufactured capital in aggregate production is inappropriate in advanced industrial countries) that sweep across regions, nations, and continents; and about those matters (such as, for example, the decline in firewood, or water availability) that are specific to the needs and concerns of poor people of as small a group as a village community. This tension should be borne in mind, and we will elaborate upon an aspect of it in the following section, when we come to evaluate an empirically based suggestion by the World Bank (1992) concerning the nature of a possible trade-off faced by poor countries between national income per head and environmental quality.

Environmental problems present themselves differently to different people. In part it is a reflection of the tension we are speaking of here. Some people identify environmental problems with wrong sorts of economic growth, while others view them through the spectacles of poverty. We will argue that both visions are correct: there is no single environmental problem; rather, there is a large collection of them. Thus, for example, growth in industrial wastes has been allied to increased economic activity, and in industrialized countries (especially those in the former Socialist block) neither preventive nor curative measures have kept pace with their production. These observations loom large not only in environmental economics, but also in the more general writings of environmentalists in the West.

On the other hand, economic growth itself has brought with it improvements in the quality of a number of environmental resources. For example, the large-scale availability of potable water, and the increased protection of human populations against both water- and airborne diseases in industrial countries, have in large measure come in the wake of the growth in national income that these countries have enjoyed over the past 200 years or so. Moreover, the physical environment inside the home has improved beyond measure with economic growth. (Cooking in South Asia continues to be a central route to respiratory illnesses among women.) Such positive links between wealth and environmental quality have not been much noted by environmental economists, nor by environmentalists in general. We would guess that this lacuna is yet another reflection of the fact that it is all too easy to overlook the enormous heterogeneity of the earth's natural consumption and capital base, ranging as it does from landscapes of scenic beauty to watering holes and sources of fuelwood. This heterogeneity should constantly be kept in mind.

5. POPULATION AND THE LOCAL RESOURCE BASE

We argued earlier that in poor countries poverty and institutional failures together are the central cause of degradation of the local environmental resource-base. This point of view is in marked contrast to the causal mechanism that has been identified in the environmental literature, which sees rapid population growth as *the* threat to the environment (e.g. Ehrlich and Ehrlich 1992). The question is: are the two explanations in conflict? We will argue that they are not, that rapid population growth is itself a result of poverty and institutional failures. High fertility rates in, say, the Indian subcontinent and Sub-Saharan Africa are a symptom of the economic malaise of these regions. In this section we will outline the arguments underlying this claim.

The spring of action pertaining to fertility matters is the household. So the right question to ask is why rural households in poor countries (for instance, those in Sub-Saharan Africa and the Indian subcontinent) pursue high fertility rates. (The total fertility rate in the Indian subcontinent is approximately 4.5 and in Sub-Saharan Africa it is about 7.) In particular, why do they not see that continued high fertility rates endanger not only their own well-being, but also that of their own descendants?

We will argue below that households in these regions have a need for large numbers of children, so that high fertility rates are a strategic response to their economic environment. (The general argument that links population, poverty, and the local environmental resource base has been explored in Dasgupta 1993.) On the other hand, it can be argued that even though individual households may act rationally in pursuing a pro-natalist concern, there is failure at the collective level. A broad reason behind collective failure can be summed up in the expression: population externalities. They occur when parents, in deciding to have an additional child, do not take into account the impact on others of their decision. A number of reasons for the occurrence of such externalities have been identified in the recent literature (see e.g. Dasgupta 1993: ch. 12) and they can be traced to both poverty and institutional failure.

Several of the chapters in this volume discuss the factors that determine a community's fertility rate. They reiterate the modern finding that these factors consist not only in the extent to which households know and have access to modern birth-control techniques, but also in the need households have for children. In short, the net *demand* for children is an important factor underlying a community's fertility rate. As we have just noted, a population externality exists when a couple's demand for an additional child is based solely on its impact on the couple, and not on its overall impact on society (that is, inclusive of its impact on other households). To be sure, we are stating the matter somewhat inaccurately when referring to a 'couple's demand' for children. The agency roles assumed by the genders are highly asymmetrical in these regions, and a separate treatment is required for identifying the effect on fertility rates of such gender biases. In this chapter we will neglect these matters for the sake of brevity. (But see Dasgupta 1993; and Bledsoe, Chapter 5 of this volume.)

Population externalities can be either beneficial (positive) or harmful (negative). In small communities, where the local resource base has not yet been severely eroded, there can be a presumption of a positive externality, which means that the community as a whole benefits when its size increases. This has been argued by a number of authors (e.g. Simon 1981), but without attention to the conditions under which positive externalities may be presumed to exist.

The main reason why there may be positive externalities when a community is richly endowed with resources is the existence of fixed costs in the organization of production and exchange of certain types of goods (for example, infrastructures), services, and ideas. The unit cost of such commodities declines when population increases, and is the source of the externality. In addition, increased possibilities of specialization in production provide yet another reason why the well-being of a society may increase when its population increases.

As against this are considerations that work in the reverse direction. The local environmental resource base provides an important instance of them. Increased population size imposes pressures on the base, in manners that have been identified in the literature on ecological economics. This in turn means the future of the community is compromised.

However, so long as the community's size is small, the benefits from growth in size would be likely to dominate its costs. But when the community has grown sufficiently large, the costs of population growth will have caught up with its benefits. As we have noted, there are many reasons for this, but an overarching one is an excessive use of the local environmental resource base. When population size is sufficiently large, its demand for the services the base provides can be expected to exceed its regeneration rate. When this occurs, the population externality can be expected to be negative (see the discussion on resilience in Section 1 above). Furthermore, the higher the population size relative to the resource base, the more urgent is the need for population control. The reason can be traced to the notion of limited resilience, with the attendant risk that the resource-base will experience irreversible changes.

Soil erosion through population pressure offers an example. When soil is eroded, farmers are forced to substitute commercial fertilizers in order to maintain agricultural output (Hyams 1976; Turner *et al.* 1990). Such an externality is not detectable in statistical information concerning agricultural output. There is simply no substitute for direct information on changes in the size and quality of the resource base if we are to assess trends in the well-being of a community. In Section 7 we will pursue this idea further when developing the concept of net national product.

6. GLOBAL EXTERNALITIES

Economic externalities take on a different complexion at the global level. Global resource consumption is dominated by those countries that are referred to as the 'North'. Economic growth (conventionally measured; see Section 7 below) comes allied to an increase in the overall demand for resources (for example, the atmosphere as a sink for emissions). This is, however, not an externality; that is, the high resource consumption in the North compared to the South does not by itself result in inefficient uses of the resources, as long as all consequences of the resource consumption are 'internalized', that is, taken into account by the decision-makers. However, in this context, the high resource consumption in the North raises three important issues:

- all consequences from resource use may not be internalized;
- an addition to the human population in the North will increase the overall resource consumption; and
- equity.

The possibility of global warming offers the clearest context in which to discuss the first of these issues. That increased emissions of carbon dioxide, among other gases, are a source of global externalities has been much discussed in recent years. What has been less publicized is that the magnitude of the current externality is shaped largely by wrong relative prices in the North. This is an example of institutional failure. It has interesting implications. If, for example, the gasoline price in the United States were to be increased through the use of a tax to levels that prevail in Europe, there would not only be a reduction in global externality, it would also be beneficial to the United States. This is an instance of what the World Bank refers to as a 'win-win policy' (see World Bank 1992). The gasoline tax would, of course, correct for the global externality (externality taxes are often referred to as Pigouvian taxes). At the same time, the resultant tax revenue would enable the US government to reduce the overall burden of domestic taxation on labour and capital. Via a computation on this mode of reasoning, it has been argued that, if a carbon tax were introduced so as to stabilize carbon emissions in the United States at the 1990 level, there would be an increase in the growth of US gross national product (see Jorgensen and Wilcoxen 1993). When institutional failures abound, there is a vast scope for exploiting win-win situations.

From a global point of view, the net gains and costs from these externalities will vary vastly among countries. Therefore, in order to eliminate the inefficiencies rising from these externalities, new global institutions will be needed with the ability to compensate those countries that will lose from increased global efficiency. These transfers can potentially be very substantial.

The second of the issues above means that a new-born individual in the North can be expected to increase the overall resource consumption on the globe, which, in turn, will increase the resource prices and thereby reduce well-being for others. The externality arises because the parents do not take this into account when deciding on the number of offspring they want. Notice that this effect exists independently of whether population is growing or not in the North. It arises whenever a family decides to have a new baby and the parents have not taken all the consequences of that decision into account. The size of the externality is obviously determined by the expected resource consumption of the new individual and the number of new individuals. The externality from one new person is greater in the North than in the South. However, the population growth in the South is much higher, which reduces in most cases the absolute differences, but does not eliminate them.

The third of the issues mentioned above is obviously the most important one. Is the present distribution of resource consumption equitable?
Most people would agree that it is far from equitable. However, to a very
large extent, the responsibility of changing this distribution belongs to
the poor countries. Unfortunately, it seems too far-fetched to suppose that
the rich countries will voluntarily reduce their consumption. On the other
hand, poor countries can, by their own decisions, promote their economic
development and thereby increase their share of resource consumption.
The aid that is now given to poor countries can probably become more
effective if it is directed towards an economic development that is consistent with sound management of the environmental resource base.

7. PROJECT EVALUATION AND THE MEASUREMENT OF NET NATIONAL PRODUCT

Whatever else it may measure, gross national product (GNP) says nothing about the use of a country's natural capital base. For that one needs
the idea of net national product (NNP), suitably measured. We have
alluded to this in earlier sections. We turn to it now.

There are two ways of assessing changes in the well-being of a society.
One would be to measure the value of changes in the constituents of well-
being (for example, the economic welfare of individuals and the
freedoms they enjoy), and the other would be to measure the value of the
alterations in the commodity determinants of well-being (that is, goods
and services that are inputs in the production of well-being). The former
procedure measures the value of alterations in various welfare measures
(for example, indices of health, education, and other social indicators),
and the latter evaluates the aggregate value of changes in the 'inputs' of
the production of well-being (for example, number of hospital beds,
schoolteachers per 1,000,000 people, indices of food expenditure and
personal consumption). A key result in modern resource allocation theory is that, provided certain technical restrictions are met, for any
conception of social well-being, and for any set of technological, transaction, information, and ecological constraints, there exists a set of
shadow (or accounting) prices of goods and services that can be used in
the estimation of real NNP. This index of NNP has the following property: small investment projects that improve the index are at once those

that increase social well-being. Here 'projects' must be interpreted very broadly, so as to encompass all the economic changes that are taking place during a time period, including changes in economic, environmental, or other policies.

We may state the matter more generally: provided the set of accounting prices is unaffected, an improvement in the index of real national product owing to an alteration in economic activities reflects an increase in social well-being. This is the sense in which real national income measures social well-being. Moreover, the sense persists no matter what is the basis upon which social well-being is founded. In particular, the use of national income in measuring changes in social well-being is *not* restricted to any particular view of what constitutes a good quality of life.

To be sure, if real national income is to reflect social well-being, accounting prices (or shadow prices) should be used. The accounting price of a commodity of service reflects its social worth. It may or may not equal the good's market price. (Accounting prices are the differences between market prices and optimum taxes and subsidies.) For environmental resources the gap between market and accounting prices is typically large, in that, because of an absence of well-defined property rights to such resources, market prices are often zero, even while their accounting prices are positive. This provides us with the sense in which it is important for poor countries to 'get their prices right': this is to bring their market prices more in line with accounting prices.

We are talking here of real *net* national product. The accounting value of the depreciation of fixed capital (and by this we mean both manufactured and natural capital) needs to be deducted if the index of national product is to play the role we are assigning to it here (see Dasgupta and Heal 1979; Hartwick 1990; Dasgupta and Mäler 1991, 1993; Mäler 1991; Lutz 1993). Thus NNP in a closed economy, when correctly measured, reads as follows:

> NNP = *Consumption + net investment in physical capital + the value of the net change in human capital + the value of the net change in the stock of natural capital − the value of current environmental damages.*[16]

It is useful to note here that the convention of regarding expenditures on public health and education as part of final demand implicitly equates the cost of their provision with the contribution they make to social well-being. This in all probability results in an underestimate in poor countries. We should note as well that current defensive expenditure against damages to the flow of environmental amenities ought to be induced in the estimation of final demand. Similarly, investment in the stock of environmental defensive capital should be included in NNP.

By investment we mean the value of net changes in capital assets, and not changes in the value of these assets. This means that anticipated capital gains (or losses) should not be included in NNP. As an example, the value of the *net* decrease in the stock of oil and natural gas (net of new discoveries, that is) ought to be deducted from GNP when NNP is estimated. The answer to the question as to how we should estimate NNP should not be a matter of opinion today: it is a matter of fact.

Current estimates of NNP are biased because depreciation of environmental resources is not deducted from GNP. Stated another way, NNP estimates are biased because a biased set of prices is in use. As we noted earlier in this section, prices imputed to environmental resources on site are usually zero. This amounts to regarding the depreciation of environmental capital as zero. But, as we also noted earlier, these resources are scarce goods, so their accounting prices are positive: they have social worth—in fact they carry in most cases a very great social worth. Profits attributed to projects that degrade the environment are, therefore, higher than the social profits they generate. This means in turn that wrong sets of projects get chosen—in both the private and public sectors.

NOTES

1. The natural capital base includes in addition minerals and other non-renewable resources.
2. International figures, such as these, are known to contain large margins of error. Nevertheless, they offer orders of magnitude. For this reason, we will allude to them. However, we will not make use of them for any other purpose.
3. There were exceptions of course (e.g. CSE, 1982, 1985; Dasgupta 1982). Moreover, agricultural and fisheries economists have routinely studied environmental matters. In the text, we are referring to a neglect of environmental matters in what could be called 'official' development economics.
4. Minerals and fossil fuels are not renewable (they are a prime example of exhaustible resources), but they raise a different set of issues. For an account of what resource-allocation theory looks like when we include exhaustible

resources in the production process, see Dasgupta and Heal (1979), Hartwick and Olewiler (1986), and Tietenberg (1988). For a non-technical account of the theory and the historical role that has been played by the substitution of new energy resources for old, see Dasgupta (1989).

5. The economic issues arising from such non-linearities are analysed in Dasgupta (1982) and Dasgupta and Mäler (1993).

6. Ehrlich and Roughgarden (1987) is an excellent treatise on these matters.

7. Anderson (1987) contains an authoritative case study of this.

8. One notable, and controversial, estimate of world-wide productivity declines in livestock and agriculture in the drylands due to soil losses was offered in UNEP (1984). The figure was an annual loss of $26bn. For a discussion of the UNEP estimate, see Gigengack *et al.* (1990). The estimate by Mabbut (1984), that approximately 40% of the productive drylands of the world are currently under threat from desertification, probably gives an idea of the magnitude of the problem. For accounts of the economics and ecology of drylands, see Falloux and Mukendi (1988) and Dixon, James, and Sherman (1989, 1990). We will discuss the notion of environmental stress in Section 3.

9. See Dasgupta (1993) for further discussion of these linkages.

10. See e.g. Falkenmark (1986, 1989), Olsen (1987), Nelson (1988), Reij, Mulder, and Begemann (1988), and Falkenmark and Chapman (1989).

11. There are added complications, among which is that the impact on the rate of regeneration of environmental resources of a wide variety of investment decisions is not fully reversible, and in some cases is quite irreversible. The capital-theoretic approach guides the exposition in Clark (1976), who, however, concentrates on fisheries. See Dasgupta (1982) for a unified capital-theoretic treatment of environmental management problems in the context of poor countries.

12. The computation is taken from Payne (1985: 7). An original source of this kind of calculations is Leach (1975), who provided estimates of energy inputs and outputs per hectare for a variety of agricultural systems (see also Bayliss Smith 1981). Higgins *et al.* (1982) estimated the carrying capacity of different types of land in poor countries. This section is taken from Dasgupta (1993).

13. 1 kcal (kilocalorie) equals approximately 4.184 kJ (kilojoules).

14. The carrying capacity of land can be increased enormously by suitable investment. In irrigated rice cultivation a hectare can support as many as 15 people, provided fertilizers, pest controls, and improved seeds are used. See Norse (1985) for sample calculations of this sort. It may be noted that China averages only about 0.09 hectares of arable land per head, while Indonesia, India and the United States average 0.12, 0.20, and 0.55, respectively (see Clark 1989).

15. They must possess some structural stability, otherwise they would have been destroyed a long time ago. The much-discussed Gaia hypothesis concerns the structural stability of the global ecosystem (see e.g. Ehrlich and Roughgarden 1987).

16. This is true only if the underlying technology is convex and continuous. We have in earlier sections of this chapter given several examples of when this assumption is not true. However, for the present purpose, we do not have to go into the technical details about how one would modify the definition of NNP because of this.

REFERENCES

Anand, S., and Ravallion, M. (1993), 'Human Development in Poor Countries: On the Role of Private Incomes and Public Services', *Journal of Economic Perspectives*, 7.

Anderson, D. (1987), *The Economics of Afforestation* (Baltimore).

Barghouti, S., and Lallement, D. (1988), 'Water Management: Problems and Potentials in the Sahelian and Sudanian Zones', in Falloux and Mukendi (1988).

Baumol, W. M., and Oates, W. (1975), *The Theory of Environmental Policy* (Englewood Cliffs, NJ).

Bayliss Smith, T. (1981), 'Seasonality and Labour in the Rural Energy Balance', in R. Chambers, R. Longhurst, and A. Pacey (eds.), *Seasonal Dimensions to Rural Poverty* (London).

Binswanger, H. (1989), 'Brazilian Policies that Encourage Deforestation in the Amazon', World Bank, Environment Department, Paper No. 16.

Binswanger, H., and Rosenzweig, M. (1986), 'Credit Markets, Wealth and Endowments in Rural South India', World Bank, Agriculture and Rural Development Department, Report No. 59.

Biswas, M., and Pinstrup-Andersen, P. (1985) (eds.), *Nutrition and Development* (Oxford).

Clark, C. W. (1976), *Mathematical Bioeconomics: The Optimal Management of Renewable Resources* (New York).

Clark, W. C. (1989), 'Managing Planet Earth', *Scientific American*, 261.

CSE (1982, 1985), *The State of India's Environment: A Citizens' Report* (New Delhi).

CSE (1990), *Human–Nature Interactions in a Central Himalayan Village: A Case Study of Village Bemru* (New Delhi).

Dasgupta, P. (1982), *The Control of Resources* (Oxford).

Dasgupta, P. (1989), 'Exhaustible Resources', in L. Friday and R. Laskey (eds.),

The Fragile Environment (Cambridge).

Dasgupta, P. (1993), *An Inquiry into Well-Being and Destitution* (Oxford).

Dasgupta, P., and Heal, G. M. (1979), *Economic Theory and Exhaustible Resources* (Cambridge).

Dasgupta, P., and Mäler, K.-G. (1991), 'The Environment and Emerging Development Issues', *Proceedings of the Annual World Bank Conference on Development Economics* (Supplement to the *World Bank Economic Review* and the *World Bank Research Observer*).

Dasgupta, P., and Mäler, K.-G. (1993), 'Poverty, Institutions and the Environmental-Resource Base', in J. Behrman and T. N. Srinivasan (eds.), *Handbook of Development Economics,* iii (Amsterdam).

de Groot, R. S. (1992), *Functions of Nature* (Amsterdam).

Dixon, J. A., James, D. E., and Sherman, P. B. (1989), *The Economics of Dryland Management* (London).

Dixon, J. A., James, D. E., and Sherman, P. B. (1990) (eds.), *Dryland Management: Economic Case Studies* (London).

Ehrlich, P. R., and Ehrlich, A. E. (1992), 'The Value of Biodiversity', *Ambio*, 21.

Ehrlich, P. R., and Roughgarden, J. (1987), *The Science of Ecology* (New York)

Ehrlich, P. R., Ehrlich, A. E., and Holdren, J. (1977), *Ecoscience: Population, Resources and the Environment* (San Francisco).

Falconer, J. (1990), *The Major Significance of 'Minor' Forest Products* (Rome: FAO)

Falconer, J., and Arnold, J. E. M. (1989), *Household Food Security and Forestry: An Analysis of Socio-Economic Issues* (Rome: FAO).

Falkenmark, M. (1986), 'Fresh Water: Time for a Modified Approach', *Ambio*, 15.

Falkenmark, M. (1989), 'The Massive Water Scarcity now Facing Africa: Why Isn't It Being Addressed?', *Ambio*, 18.

Falkenmark, M., and Chapman, T. (1989) (eds.), *Comparative Hydrology: An Ecological Approach to Land and Water Resources* (Paris: UNESCO).

Falloux, F., and Mukendi, A. (1988) (eds.), 'Desertification Control and Renewable Resource Management in the Sahelian and Sudanian Zones of West Africa', World Bank, Technical Paper No. 70.

Feder, E. (1977), 'Agribusiness and the Elimination of Latin America's Rural Proletariat', *World Development*, 5.

Folke, C. (1991), 'Socioeconomic Dependence of the Life-Supporting Environment', in C. Folke and T. Kåberger (eds.), *Linking the Natural Environment and the Economy: Essays from the Eco-Eco Group* (Dordrecht).

Folke, C., and Kautsky, N. (1992), 'Aquaculture with its Environment: Prospects for Sustainability', *Ocean and Coastal Management,* 17.

Gigengack, A. R. *et al.* (1990), 'Global Modelling of Dryland Degradation', in

Dixon, James, and Sherman (1990).

Gross, D. R., and Underwood, B. A. (1971), 'Technological Change and Caloric Cost: Sisal Agriculture in North-Eastern Brazil', *American Anthropologist*, 73.

Hamilton, L. S., and King, P. N. (1983), *Tropical Forested Watersheds: Hydrologic and Soils Response to Major Uses or Conversions* (Boulder, Colo.).

Hartwick, J. (1990), 'Natural Resource, National Accounting, and Economic Depreciation', *Journal of Public Economics*, 43.

Hartwick, J., and Olewiler, N. (1986), *The Economics of Natural Resource Use* (New York).

Higgins, G. M., *et al.* (1982), *Potential Population Supporting Capacities of Lands in Developing Countries* (Rome: FAO).

Hoff, K., and Stiglitz, J. E. (1990), 'Introduction: Imperfect Information and Rural Credit Markets: Puzzles and Policy Perspectives', *World Bank Economic Review*, 4.

Holling, C. S., Schindler, D. W., Walker, B. W., and Roughgarden, J. (1994), 'Biodiversity in the Functioning of Ecosystems: An Ecological Primer and Synthesis', in Perrings *et al.* (1994).

Hyams, Edward, S. (1976), *Soil and Civilization*, (New York).

Jorgensen D., and Wilcoxen, P. (1993), 'Reducing US Carbon Emissions: An Ecometric General Equilibrium Assessment', *Resource and Energy Economics*, 15/1.

Kennedy, E., and Oniang'o, R. (1990), 'Health and Nutrition Effects of Sugarcane Production in South-Western Kenya', *Food and Nutrition Bulletin*, 12.

Leach, G. (1975), 'Energy and Food Production', *Food Policy*, 1.

Lutz, E., (1993) (ed.), *Toward Improved Accounting for the Environment* (Washington, DC: World Bank).

Mabbut, J. (1984), 'A New Global Assessment of the Status and Trends of Desertification', *Environmental Conservation*, 11.

Mahar, D. (1988), 'Government Policies and Deforestation in Brazil's Amazon Region', World Bank, Environment Department, Paper No. 7.

Mäler, K.-G. (1974), *Environmental Economics: A Theoretical Enquiry* (Baltimore).

Mäler, K.-G. (1991), 'National Accounting and Environmental Resources', *Journal of Environmental Economics and Resources*, 1.

Meade, J. E. (1973), *The Theory of Externalities* (Geneva).

Musgrave, R. (1959), *Theory of Public Finance* (New York).

Nelson, R. (1988), 'Dryland Management: The "Desertification" Problem', World Bank, Environment Department, Working Paper No. 8.

Norse, D. (1985), 'Nutritional Implications of Resource Policies and Technological Change', in Biswas and Pinstrup-Andersen (1985).

Odum, E. P. (1975), *Ecology* (New York).

Olsen, W. K. (1987), 'Manmade "Drought" in Rayalaseema', *Economic and Political Weekly*, 22.

Payne, P. (1985), 'The Nature of Malnutrition', in Biswas and Pinstrup-Andersen (1985).

Perrings, C., and Walker, B. W. (1994), 'Biodiversity Loss and the Economics of Discontinuous Change in Semi-Arid Rangelands', in Perrings *et al.* (1994).

Perrings, C., Mäler, K.-G., Folke, C., Holling, C. S., and Jansson, B.-O. (1994) (eds.), *Biodiversity Loss: Ecological and Economic Issues* (Cambridge).

Penman, H. L. (1956), 'Evaporation: An Introductory Survey', *Netherlands Journal of Agricultural Science*.

Pigou, A. C. (1920), *The Economics of Welfare* (London).

Portes, R. (1971), 'Decentralized Planning Procedures and Centrally Planned Economies', *American Economics Review* (Papers and Proceedings), 61.

Pimentel, D., and Pimentel, M. (1979), *Food, Energy, and Society* (New York).

Regier, H. A., and Baskerville, G. L. (1986), 'Sustainable Redevelopment of Regional Ecosystems Degraded by Exploitive Development', in E. C. Clark and R. E. Munn (eds.), *Sustainable Development of the Biosphere* (Cambridge).

Reij, C., Mulder, P., and Begemann, L. (1988), 'Water Harvesting for Plant Production', World Bank, Technical Paper No. 91.

Repetto, R. (1988), 'Economic Policy Reform for Natural Resource Conservation', World Bank, Environment Department, Working Paper No. 4.

Rosenzweig, M., and Wolpin, K. I. (1985), 'Specific Experience, Household Structure and Intergenerational Transfers: Farm Family Land and Labour Arrangements in Developing Countries', *Quarterly Journal of Economics*, 100.

Sen, A. (1992), *Inequality Re-examined* (Oxford).

Simon, J. (1981), *The Ultimate Resource* (Princeton, NJ).

Solbrig, O. T. (1993), 'Plant Traits and Adaptive Strategies: Their Role in Ecosystem Function', in E.-D. Schulze and H. A. Mooney (eds.), *Biodiversity and Ecosystem Function* (Heidelberg).

Solorzano, R., *et al.* (1991), *Accounts Overdue: Natural Resource Depreciation in Costa Rica* (Washington, DC: World Resources Institute).

Tietenberg, T. (1988), *Environmental and Natural Resource Economics*, 2nd edn. (Glenview, Ill.).

Trenbath, B. R., Conway, G. R., and Craig, I. A. (1990), 'Threats to Sustainability in Intensified Agricultural Systems: Analysis and Implications for Management', in S. R. Gliessman (ed.), *Agroecology* (New York).

Turner, B. L., et al. (1990), *The Earth as Transformed by Human Action*, (Cambridge).

UNEP (1984), *General Assessment of Progress in the Implementation of the Plan of Action to Combat Desertification 1978-1984*, Report of the Executive Director (Nairobi: United Nations Environment Programme)

United Nations (1990), *Overall Socioeconomic Perspectives of the World Economy to the Year 2000* (New York: UN Department of International Economic and Social Affairs).

Walters, C. J. (1986), *Adaptive Management of Renewable Resources* (New York).

von Braun, J., and Kennedy, E. (1986), 'Commercialization of Subsistence Agriculture: Income and Nutritional Effects in Developing Countries', International Food Research Institute, Washington DC, Working Paper on Commercialization of Agriculture and Nutrition No. 1.

Wilson, E. O. (1992), *The Diversity of Life* (New York).

World Bank (1992), *World Development Report* (New York).

3

Population and Reasoned Agency: Food, Fertility, and Economic Development

Amartya Sen[*]

1. THE ISSUES

'Mr. Condorcet's picture of what may be expected to happen when the number of men shall surpass the means of their subsistence is justly drawn,' said Thomas Malthus in 1798.[1] He was commenting on an analysis of Condorcet, published a few years earlier, that we now associate with Malthus himself. After quoting Condorcet with approval, Malthus proceeded to express his emendations, in some important ways, of Condorcet's approach, in particular, replacing Condorcet's optimism (founded on his belief in the human ability to make intelligent use of science and to take rational decisions to deal with identified problems).

With the revival, in various modern forms, of Malthusian pessimism and distrust of human rationality, the differences that separated Condorcet and Malthus have become particularly relevant now. While, on the nature and tempestuousness of the population problem, I have some differences with Paul Kennedy, whose forceful presentation of one type of Malthusian worry has been extremely influential in public discussions, I do agree very much with him that 'this debate between optimists [Godwin, Condorcet] and pessimists [Malthus] has, in one form or another, been with us since then', and that 'it is even more pertinent today than when Malthus composed his *Essay*'.[2]

* Lamont University Professor, Harvard University. For helpful discussions, I am most grateful to the participants of the Population–Environment–Development Seminars at the Royal Swedish Academy of Sciences and the Beijer Institute, on 11 November 1993, and 7–8 January 1994. For useful comments and suggestions, I am also indebted to Lincoln Chen, Partha Dasgupta, Jan Lundqvist, Karl-Göran Mäler, Gerald Piel, Emma Rothschild, Sanjay Reddy, Vernon Ruttan, Peter Timmer, and Jeffrey Williamson.

What was the debate about? Malthus quoted Condorcet fairly extensively, and in particular focused on the following passage from
Condorcet (1795: 256-7):

But in this progress of industry and happiness, each generation will be called to
more extended enjoyments, and in consequence, by the physical constitution of
human frame, to an increase in the number of individuals. Must not there arrive
a period then, when these laws, equally necessary, shall counteract each other?
When the increase of the number of men surpassing their means of subsistence,
the necessary result must be either a continual diminution of happiness and
population, a movement truly retrograde, or, at least, a kind of oscillation
between good and evil? In societies arrived at this term, will not this oscillation
be a constantly subsisting cause of periodical misery? (English translation from
Malthus 1798: 123)

If these rhetorical questions seem a little Malthusian in spirit (though
penned, earlier, by Condorcet), it will not come as a total surprise that
Malthus expressed his full agreement with all this: 'The oscillation
which he describes will certainly take place and will without doubt be a
constantly subsisting cause of periodical misery.' But Condorcet, while
in favour of asking these questions, had gone on to indicate how a human
society based on science and rationality will overcome these problems.
Malthus disagreed as firmly with Condorcet's answers as he had agreed
with Condorcet's posing of these questions (pointing to the *possibility* of
these problems).

Where exactly did they differ? I shall focus on three contrasts here—
relevant to the population–development–environment debate (though
there were also other differences between the two).

(1) *How close are we to the limits?* The first difference relates to
Condorcet's belief that the population at his time was not anywhere near
any eventual upper limit that might exist. This is a point on which
Malthus thoroughly disagreed. He saw the stable limits of population
size as having been crossed in his own time, and indeed much earlier, and
argued that mankind was firmly caught *already* in the oscillation of
which Condorcet had spoken:

The only point in which I differ from Mr. Condorcet with regard to this picture is
the period when it may be applied to the human race. Mr. Condorcet thinks that
it cannot possibly be applicable but at an era extremely distant. If the proportion
between the natural increase of population and food which I have given be in any

degree near the truth, it will appear, on the contrary, that the period when the number of men surpass their means of subsistence has long since arrived, and that this necessary oscillation, this constantly subsisting cause of periodical misery, has existed ever since we have had any histories of mankind, does exist at present, and will for ever continue to exist, unless some decided change take place in the physical construction of our nature. (Malthus 1798: 123–4)

The history of massive population growth combined with continued increase in living standards since the late eighteenth century has not given much comfort to Malthus's rejection of Condorcet's less pessimistic prognosis. Nevertheless, the question 'How near are the limits?' can be fruitfully re-examined now taking note of the current situation and the contemporary empirical realities (as we, to quote Paul Kennedy's appropriate phrase, 'prepare for the twenty-first century').

(2) *Is food the main problem?* In the passage that Malthus had quoted about the possibility of population surpassing our means, Condorcet had not identified *food* as being the specific problem related to overpopulation. However, Malthus's formulation of Condorcet's concern reformulated all this directly into the very particular issue of 'the proportion between the natural increase of *population and food*'.[3] Condorcet does, of course, discuss, here as well as elsewhere, the food problem (among several other problems), but his basic interest lies much more generally in the quality of life (and even in 'the perfectibility' of human beings), and he is specifically concerned here with critically examining the possibility of a decline in 'the happiness of the human race' as a result of population increase.

Again, while the world has become a much more crowded place since the times of Condorcet and Malthus, and many particular problems associated with that expansion—including environmental threats—can be identified, there has been little evidence that the production of food, as such, has been falling behind the growth of world population. However, we must still ask, two centuries beyond Condorcet and Malthus, whether food production may not—*now*—have become (or be in the process of becoming) the right focus for worry about population growth in the world in which we live.

(3) *Can a rational social policy be voluntary?* Condorcet saw the problem of population size as being resolvable through reasoned understanding of the consequences of population growth, and intelligent use of scientific knowledge, social arrangements, and rational individual

behaviour. He did see the need for correcting distortions of incentives, and argued for making individual incentives come more in line with ethical concerns (1795: 192): 'Will not a country's constitution and laws accord best with the rights of reason and nature when the path of virtue is no longer arduous and when the temptations that lead men from it are few and feeble?'

Condorcet also took a broad view of personal 'interest' based on enlightenment and reason (1795: 192): 'Is not a mistaken sense of interest the most common cause of actions contrary to the general welfare? Is not the violence of our passions often the result either of habits that we have adopted through miscalculation, or of our ignorance how to restrain them, tame them, deflect them, rule them?' The nature and reach of reasoned incentives, I would argue, are among the major issues today in assessing the rival claims of compulsion and voluntarism—not least in dealing with the population problem.

On the production side, Condorcet talked not only about general technical progress (even of 'a degree of knowledge of which we can have no inkling'), but also specifically of using 'methods of preservation and economy in expenditure' that will also improve with 'progress in the arts of producing and preparing supplies and making articles from them'.[4] On the side of population growth, Condorcet anticipated the emergence of new norms of family size based on 'the progress of reason'. He predicted a time when 'the absurd prejudices of superstition will have ceased to corrupt and degrade the moral code by its harsh doctrines', and when people 'will know that, if they have a duty towards those who are not yet born, that duty is not to give them existence but to give them happiness', so that they will on their own choose to restrain family size, 'rather than foolishly to encumber the world with useless and wretched beings' (1795: 188–9).

Malthus, on the other hand, saw little chance of solving social problems through reasoned decisions, took the limits of food production to be relatively inflexible, and was particularly sceptical of voluntary family planning. He did talk of 'moral restraint' as an alternative way of reducing the pressure of population (alternative, that is, to misery and 'positive checks'), and varied a little over the years on the effectiveness of this course; but on the whole he saw little prospect of success. He took 'preventive checks' as working mainly through 'the restraint from marriage', and returned persistently to the view that 'there is no reason whatever to suppose that anything beside the difficulty of procuring in adequate plenty the necessaries of life should either indispose this greater number of persons to marry early, or disable them from rearing in health the largest families' (1830: 243). Not only did this lead Malthus to oppose public relief of poverty;[5] it also left him deeply pessimistic about the realistic possibility of solving the population problem through rational decisions.

What about the use of compulsion in restricting families—issues that are much in focus now, particularly stimulated by the apparently successful Chinese policies of forceful family restriction? Condorcet had seen no need for force and compulsion, given his belief in the use of science, social reform, and the individual capacity to respond rationally to identified problems and reasoned incentives.

Malthus also, for rather different reasons, did not advocate compulsion in population control. Rather, he accepted the necessity of periodic miseries, and saw some merit in it (1798: 206, 209): 'It seems, however, every way probable that even the acknowledged difficulties occasioned by the law of population tend rather to promote than impede the general purpose of Providence'. ' The general tendency of an uniform course of prosperity is rather to degrade than exalt the character.' However, for a contemporary social thinker, less enamoured of improving humanity (and fulfilling the purpose of Providence) through pain and misery, it would be natural to move from Malthus's pessimism about the voluntary solution of the population problem to advocate social coercion in the cause of reducing family size. And that extension has been among the forms that the recent revivals of Malthusian perspectives have tended to take in contemporary discussions.

2. FOOD OUTPUT, POPULATION, AND GEOGRAPHY

How has food production fared compared with population growth since Malthus's time? Between the time when Malthus wrote his *Essay* and now the world population has grown about six times, and yet food production has kept up—and more—with population growth without much sign of strain. But this is too long a run to tell us very much about where we stand today. The world population is now above 5.5 billion (it was less than a billion in Malthus's days); it grew by 923 million during the

decade 1980–90; and—as is often pointed out in the popular media—
about 90 per cent of that increase occurred in the developing countries
(World Population News Service 1993: 1). (See Table 3.1 for a breakdown of population expansion among regions.) We can well ask, in these
new circumstances, how food production is doing compared with the
contemporary growth of population.

TABLE 3.1 *Population growth in different regions of the world, 1980–1990*

Regions	Population (millions)		Decadal growth	
	1980	1990	Total (millions)	Annual %
World	4,428	5,351	923	1.7
Middle East and North Africa	173	244	71	3.2
Sub-Saharan Africa	351	489	138	3.1
South Asia	903	1,152	249	2.2
of which: India	684	850	166	2.1
Latin America	358	455	97	2.0
East Asia and Pacific	1,399	1,667	268	1,6
of which: China	988	1,134	146	1,5

Note: Latin America includes the Caribbean. The population annual growth figures for India and China
are for 1980–91, rather than for 1980–90

Source: Based on data presented in *World Development Report* (1993: Tables A.1, A.3, A.4, and 26).

Has the world food output started falling behind world population in
recent years? Though there are often pronouncements in that direction,
especially if food production grows by only a little in some particular
year, there is, in fact, no serious evidence whatever to indicate that the
trend of food output per head has become a downward one.[6] While there
are some year-to-year fluctuations, the trend is clear enough. For example, food production per head now (that is, for the most recent three
years, 1990 to 1992) is more than 4 per cent higher than about a decade
ago (1979–81), and more than 6 per cent higher than a decade and a half
earlier (1974–6).[7] Similar increases can be found in the preceding decades as well. As we move to the next century, it is hard to detect, despite
the frequently repeated rhetoric in the media, any serious sign whatever
that the world food output is falling behind the growth of world population.

However, we must bear in mind the fact that around 90 per cent of the
increase in population is taking place in the poorer countries in the
world—about two-thirds of it in Asia. In that context, we should ask
where exactly the increase in food production per head is occurring. Is

the upward global trend of per-capita food output the result of high performance in producing food output in the rich countries only? This is not the case. (Table 3.2 presents the changes in food production by the major regions.) For example, over the last decade, food output per head in Europe rose more slowly than in the world, and the figure for the United States recorded nearly a 3 per-cent decline. In contrast, in Asia, where two-thirds of the absolute increase in population in the developing world took place, food output per head rose by more than 20 per cent (with a growth of more than 22 per cent in India and 36 per cent in China).

TABLE 3.2 *Indices of food production per head by regions*

Regions	1974–6	1979–81	1985–7	1990–2
World	97,4	100.0	103.8	104.1
Europe	94.7	100.0	106.4	103.5
North America	90.1	100.0	99.1	99.2
of which: USA	89.8	100.0	97.9	97.2
Other developed	107.5	100.0	98.3	91.6
Africa	104.9	100.0	95.8	94.2
Latin America	94.0	100.0	100.5	103.1
Asia	94.7	100.0	111.9	120.8
of which: India	96.5	100.0	109.2	122.4
China	90.1	100.0	122.7	136.4

Source: With the three-year average of 1979–81 as the base, the three-year averages for the later years (1985–7 and 1990–2) are obtained from United Nations (1993: table 4). The three-year average for the earlier years, 1974–6, are based on United Nations (1984: table 1). There may be slight differences in the relative weights between the two sets of comparisons, so that the series should not be taken to be fully comparable between the two sides of 1979–81, but the quantitative difference made by this, if any, is likely to be quite small.

The problematic region, in this respect, has been Africa, where food output per capita fell by nearly 6 per cent over this period. More specifically, it is in Sub-Saharan Africa that such decline can be widely observed. That region, incidentally, contributed a relatively modest 15 per cent to the increase in world population over the last decade (compared with 90 per cent for the developing countries as a whole, which experienced, in general, significantly rising ratio of food output to population). But the problems faced by Sub-Saharan Africa are indeed serious, and we must examine them carefully (see Section 5 below).

At this stage, it would be useful to divide the problem of food production into two distinct aspects: (1) the general issue of balancing the expansion of food production to population growth for the world as a whole, and (2) the particular problems of specific countries and regions (particularly Sub-Saharan Africa) in balancing food supply and population size. I shall address them in turn. But, in addition, we must also pay attention to two related issues: (3) the linkage between population pressure, production possibilities, and the environment, and (4) the influences that operate on the rate of expansion of population itself (taking that opportunity to re-examine the old question of the limits of 'voluntarism').

3. OUTPUT, PRODUCTION POSSIBILITY, AND ECONOMIC INCENTIVES

Looking at food output per head is not really as enlightening as it might at first appear. Food is produced by peasants, farmers, and others not to *demonstrate* how much can be grown, but to make *economic use* of it— to eat, to sell, to exchange. We cannot directly infer how much *could* have been produced merely by looking at what was actually produced. To be sure, we do know that what was actually produced certainly was possible, but we do not know how much more could have been produced if there were economic incentives for expanding output.

This economic distinction, though extremely elementary, is quite important in assessing the global food problem. The optimists may tend to overestimate the basis of their hopefulness by simple extrapolation— that is, by just assuming that food production can be pushed up faster than population growth, as has happened in the past, without critically examining the actual possibility of expanding production that exists today. The pessimists, on the other hand, may note that food production is growing only a little faster than population, and this they may tend to interpret as evidence that we are reaching the limits of what we can produce. Such a presumption would not be right, since it ignores the effects of economic incentives that govern production: food will not be produced beyond the effective demand for it.

How do we think about what *can* be produced? One of the economic issues to bear in mind is the fact that the production of a particular commodity (*A*) can be expanded, even with a given resource base, by reducing the production of another (*B*). Typically, such a process of 'transformation' would be accompanied by commodity *A* becoming more and more expensive *vis-à-vis* commodity B, as resources are shifted into making A by being withdrawn from the production of B, and increasingly more and more resources are needed per unit of production of A *vis-à-vis* that of B.[8] Over time, we may watch two processes simultaneously: (1) expansion of the productive base due to technical progress and resource accumulation (this can expand the capacity to produce both A and B), and (2) expansion of the output of A at the cost of B (by shifting resources to 'transform' the production of B into that of A). The observation of actual production baskets over time has to be combined with interpretation of what is going on.

It is hard to arrive at definitive interpretations of the production changes, taking full note of all the complexities of technological and economic relations. It is, nevertheless, of interest to enquire whether there are some prima-facie signs that indicate that expansion of food production is being made to keep up with (and exceed) the growth of population by a strained process of 'transformation', and by absorbing more and more resources into making units of food compared with units of other commodities. A plausible sign of that would be a general increase in the relative price of food compared with other products. Do we observe such a tendency?

There is, in fact, very little evidence in that direction. If we look at the long haul, the output of non-food commodities has grown even faster—indeed very much faster—than that of food, across the world. There is also little to indicate that food has become much more expensive in terms of other commodities, which would give tell-tale signs of needing more and more resources per unit compared with other commodity production. For example, if we look at the price of wheat, corrected for general inflation (at constant US dollars), in the US markets between 1800 (around the time of Malthus's first *Essay on Population*) and the mid-1980s, we see sharp fluctuations but no general upward trend—in fact, quite the contrary (see Fig. 3.1). Wheat is cheaper today, in terms of other commodities, than it was in Malthus's time.

FIG. 3.1 Price of wheat in constant US dollars, 1800–1985

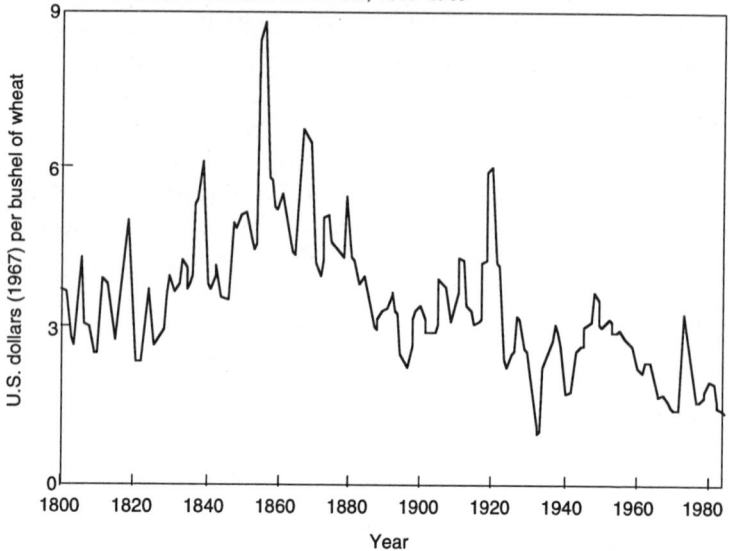

Note: The dollar values are for constant 1967 dollars
Source: Piel (1992: 223)

Has this picture been changing in recent years, and are there some signs of strained 'transformation' now? The information here is too diverse and too hard to interpret for us to arrive at any kind of a clear picture, but it is interesting to note that the world of international economic relations has been full of complaints of falling food prices *vis-à-vis* other commodities. For example, the United Nations *Report* by the Secretariat to the UNCTAD VIII Conference in 1992 records a 38 per cent fall in the 'real prices' of 'basic foods' over the last decade.[9] This picture is in line with another which gives the 'real prices' of particular food items exported by developing countries, stretching over the last *three* decades (between 1960 and 1988), adjusted by the index of prices of manufactured goods (see Table 3.3).

These indicators, as calculated by the UNCTAD, can be supplemented by the World Bank's estimates of real prices of particular food crops between 1953 and 1985, adjusted by the 'manufacturing unit value index' (see Table 3.4). There seems to have been a considerable fall in the relative price of staple food *vis-à-vis* manufactured goods, rather than the opposite.

TABLE 3.3 *Trend of international food prices* vis-à-vis *manufacturing, prices: average annual rate of change* (%)

Food	1960–71	1972–80	1981–88
Wheat	–3.0	–5.2	–10.6
Rice	–0.6	–5.3	–14.0
Maize	–0.9	–4.0	–10.4
Sugar	–5.9	–7.6	–17.3
Bananas	–2.9	–0.5	–4.2
Soybean meal	–0.1	–6.0	–7.9
Fish meal	–0.6	–5.5	–7.5
Meat (bovine)	+6.5	–3.8	–4.8

Note: International market prices have been used for the respective commodities, adjusted by the export unit values of manufacturing goods. For rice, sugar and bananas, the free-market prices differ from the controlled prices based on legal agreements, and the former prices have been used.

Source: United Nations (1989: Table 4.11).

TABLE 3.4 Food prices in constant 1985 US $: 1953–1955 to 1983–1985

Food	1953–5	1983–5	%-age change
Wheat	232.4	135.3	–41.8
Rice	572.5	248.0	–56.7
Sorghum	191.7	116.8	–39.1
Maize	202.1	127.9	–36.7

Note: The prices are international, and in the case of wheat it is the price of US wheat. The units are constant (1985) US dollars, adjusted by 'the manufacturing unit value index'.

Source: World Bank (1989: Annex Tables 6, 12, 18).

Questions can certainly be raised about the reliability as well as the interpretation of these price data. The international prices may diverge from prices and costs that obtain in particular economies. Some countries subsidize irrigation, fertilizers, and other agricultural inputs, and we have to correct for changes in them. World food prices are also distorted by various controls that affect international markets, though these interventions tend to go typically in the direction of keeping food prices up, rather than pushing them down. Also, the 'real prices' here compare prices of food with those of manufactured goods only. But it is easily checked that the prices of services have risen sharply over time compared with manufactured goods, so that their inclusion would make the fall in the 'real price' of food greater rather than less.

There remain many uncertainties in interpreting these data (the limitation of market prices as indicators of true social costs is one of them), but on the basis of these—and similar data from other sources—it would be hard to conclude that there is any presumption that food output is being made to keep up with population growth through strenuous 'transformation' exercises, making it increasingly more costly to produce food, compared with other goods and services.[10] If anything, the evidence points in the opposite direction.

4. INCENTIVES, ENTITLEMENTS, AND FOOD CONSUMPTION

In considering the future demand for food, we have, however, to take note not only of the increase in population, but also of rising food consumption per head. While population growth is slowing down, there will continue to be some expansion of food consumption per head with the growing prosperity of those who are forced now, by their poor economic circumstances, to remain somewhat hungry.

The neo-Malthusian concern with population size tends to distract attention from the importance of demand related to economic means and their consequences.[11] In fact, even famines have occurred initiated by sudden and sharp increases in the economic means of large and relatively poor sections of the population, leading to rapid rises in demand for food and its prices, making it harder for *other* poor people to buy the food that was there.[12] In considering future demands for food, these '*non*-Malthusian' issues of increased food consumption per person would have to be considered seriously, but they must not be confused with the so-called 'Malthusian' worry based simply on the number of 'mouths' to feed.

The point is sometimes made that, while the Malthusian concern with the ratio of food to population may be simple-minded, it is still a good basis for public discussions since it draws attention to an important aspect of the challenges we face today. This argument is deceptive, not only because mis-describing the fuller picture of a challenge may not help to solve it, but also because that 'Malthusian indicator' can lull policy analysts into smugness when a crisis arises that is not reflected in that particular indicator. I have discussed elsewhere several cases of famines where the governments in office failed to take preventive action largely because they did not anticipate a famine in view of the fact that the food output and availability per head were quite high (that is, precisely as a result of the so-called 'Malthusian indicator' signalling that there was no reason to anticipate starvation). (See also Sen 1981: chs. 6, 7, and 9). While Malthusian pessimism is the subject of much debate, it must be noted that what may be called 'Malthusian optimism' (fearing no famine when total food per person is high) has killed many million people.

In fact, the historical increase in food output per head that we observed (and on which I commented earlier) is largely a reflection of the growing average prosperity of the world population. It is precisely because people can afford to eat more food that the effective demand for food per person has expanded, thereby providing incentive for producing more food per head.

The neglect of the issue of economic incentives in interpreting production and trade is a problem that afflicts many of the public discussions of these and related questions. The actual production of food and its disposal depend on demand and the profitable opportunities of selling it. They are governed neither by an autonomous use of production possibility, nor by any reading of needs, irrespective of purchasing power. People die in famines even when the overall food supply may not be much reduced, or reduced at all, precisely because some do not have the economic means to command food. In fact, sometimes food even moves *out* of famine districts to elsewhere in search of a better price in non-stricken areas (as happened, for example, in the Irish famines of the 1840s, or in the Wollo famine in Ethiopia in 1973).[13]

The error in judging production possibility by simply looking at actual production (without examining the role of effective demand and incentives in determining production) is, in some ways, rather similar to the fateful error in some traditional analyses of famines and hunger that look only at the production and availability of food (without examining the role of effective demand and incentives in determining distribution).

5. WHAT ABOUT SUB-SAHARAN AFRICA?

In Sections 3 and 4 my focus has been on global aspects of the food problem. But, as was noted in Section 2, there is clear evidence that food output per head has fallen in Sub-Saharan Africa. Since starvation and famines are reported persistently from this region of the world, it might appear that here at last we have a classic case of Malthusian famine, generated by falling food output per capita.

But that would be an over-hasty conclusion. Food output per head has fallen in many economies in the world without there being any resulting famine or starvation. Substantial decline in food output per head has been experienced, among others, by countries ranging from Barbados, Cyprus, and Guyana, to Singapore, Sri Lanka, Sweden, and Syria (see Table 3.5). These countries could purchase food from elsewhere, since they had the economic means to do so. To a great extent many of the Sub-Saharan economies also did just this, and figures of calorie consumption do not show the sharp declines that are observed in the statistics of food output per capita.[14] Nevertheless, the question arises, in so far as Sub-Saharan Africa has had more famines and related deprivations than other parts of the world, of where the difference lies.

The features that distinguish many Sub-Saharan African economies from other countries which have also experienced—for shorter or longer stretches of time—considerable decline in food output per head have been discussed elsewhere (see Sen 1981, 1984; Drèze and Sen 1989; 1990). I shall briefly recapitulate here some of the main points of difference.

First, because such a high proportion of people rely on food production for earning their income in Sub-Saharan Africa, a fall in food output is typically accompanied by the collapse of individual means of purchase and thus by an inability to buy food. The role of food production in providing a supply of particular commodities has to be distinguished from its role as the main source of income of a large section of the population. Secondly, there is some evidence that the relatively more authoritarian forms of government—such as military rules—that have tended to be more common in Sub-Saharan Africa than elsewhere are much less sensitive to the plight of famine victims, and tend to be slower in taking preventive actions in helping potential famine victims with compensatory support. Indeed, the less authoritarian governments in Sub-Saharan Africa (for example, in Botswana and Zimbabwe) have often been able to avert famines through timely intervention in re-creating lost income, even when the declines in food outputs and family incomes have been very large.

Thirdly, many of the Sub-Saharan economies are too poor—and some of them too debt-ridden—to be able to buy food from abroad with ease. In fact, given the share of food production in the aggregate income of the economy, the fall in food production also entails a much larger dip in their total economic means (unlike in, say, Sweden or Singapore, or even

TABLE 3.5 *Indices of food production per head by countries*

	1974–6	1979–81	1985–7	1990–2
Africa				
Algeria	114.1	100.0	110.9	122.3
Angola	105.3	100.0	87.5	77.8
Benin	96.8	100.0	111.9	118.4
Botswana	144.9	100.0	74.8	71.9
Burkina Faso	98.0	100.0	116.3	125.3
Burundi	101.0	100.0	102.6	98.8
Cameroon	114.1	100.0	95.8	83.0
Central African Republic	103.5	100.0	93.2	92.1
Chad	94.9	100.0	98.6	102.4
Congo	100.1	100.0	97.7	94.5
Egypt	106.8	100.0	108.6	116.8
Ethiopia	96.6	100.0	90.4	85.5
Gabon	99.0	100.0	83.1	83.5
Gambia	148.5	100.0	102.1	86.1
Ghana	136.4	100.0	107.0	104.1
Guinea	115.4	100.0	96.4	98.2
Guinea Bissau	137.0	100.0	110.7	110.3
Ivory Coast	87.7	100.0	98.2	93.3
Kenya	115.4	100.0	105.6	98.8
Lesotho	119.0	100.0	79.3	73.8
Liberia	106.0	100.0	97.5	65.0
Libya	111.1	100.0	88.7	91.8
Madagascar	111.9	100.0	95.0	87.1
Malawi	101.0	100.0	83.4	60.1
Mali	92.3	100.0	95.1	92.4
Mauritania	90.9	100.0	89.8	85.6
Mauriutius	111.5	100.0	108.9	109.1
Morocco	107.1	100.0	115.9	122.0
Mozambique	130.4	100.0	92.0	83.5
Namibia	123.0	100.0	76.6	72.3
Niger	78.1	100.0	70.0	85.0
Nigeria	99.7	100.0	100.4	121.7
Reunion	92.0	100.0	76.2	79.8
Rwanda	90.1	100.0	92.3	79.9
Senegal	151.5	100.0	112.3	96.0
Sierra Leone	107.5	100.0	97.9	88.2
Somalia	132.7	100.0	*	*
South Africa	97.4	100.0	83.8	74.1
Sudan	100.0	100.0	83.7	80.9
Swaziland	92.9	100.0	102.3	89.9
Tanzania	92.3	100.0	94.0	83.2
Togo	99.0	100.0	91.7	97.5
Tunisia	111.1	100.0	114.4	122.4
Uganda	109.5	100.0	98.0	105.4
Zaire	108.3	100.0	98.1	91.9
Zambia	127.1	100.0	94.5	85.8
Zimbabwe	121.0	100.0	97.7	70.7
North America				
Canada	91.5	100.0	111.4	111.4
United States	89.8	100.0	97.9	97.2

TABLE 3.5 *Indices of food production per head by countries*

	1974–6	1979–81	1985–7	1990–2
Caribbean and Central America				
Barbados	84.3	100.0	78.2	72.9
Costa Rica	106.0	100.0	91.0	97.2
Cuba	81.7	100.0	103.8	94.0
Dominican Republic	104.9	100.0	95.9	86.2
El Salvador	100.7	100.0	94.1	105.4
Guadeloupe	112.4	100.0	109.0	86.4
Guatemala	94.0	100.0	98.0	97.1
Haiti	106.8	100.0	100.8	80.3
Honduras	94.6	100.0	86.1	92.2
Jamaica	101.4	100.0	99.4	103.1
Martinique	121.0	100.0	119.2	123.2
Mexico	92.0	100.0	98.9	99.9
Nicaragua	111.5	100.0	68.2	63.3
Panama	97.7	100.0	97.6	87.0
Puerto Rico	103.1	100.0	87.2	90.1
Trinidad and Tobago	132.7	100.0	76.1	90.9
South America				
Argentina	90.9	100.0	98.1	97.2
Bolivia	107.1	100.0	98.3	117.3
Brazil	91.7	100.0	105.3	110.6
Chile	97.7	100.0	99.6	116.0
Colombia	91.7	100.0	98.0	113.0
Ecuador	103.4	100.0	101.3	109.7
Guyana	107.9	100.0	80.4	62.3
Paraguay	90.9	100.0	109.4	102.2
Peru	118.6	100.0	99.6	93.6
Surinam	74.1	100.0	102.4	83.1
Uruguay	104.9	100.0	100.9	114.0
Venezuela	103.4	100.0	102.0	100.2
Asia				
Afghanistan	100.0	100.0	88.9	69.3
Bangladesh	98.7	100.0	95.5	95.6
Bhutan	97.4	100.0	100.1	84.7
Cambodia	119.0	100.0	131.3	139.6
China	90.1	100.0	122.7	136.4
Cyprus	90.4	100.0	83.7	74.1
Hong Kong	91.7	100.0	*	*
India	96.5	100.0	109.2	122.4
Indonesia	87.2	100.0	118.6	135.0
Iran	101.7	100.0	110.9	118.6
Iraq	98.0	100.0	112.5	85.3
Israel	110.3	100.0	112.4	109.0
Japan	107.5	100.0	102.8	93.0
Jordan	97.1	100.0	118.2	112.0
Korea DRP	89.8	100.0	106.9	103.4
Korea Rep.	90.6	100.0	101.0	102.1
Laos	85.0	100.0	115.4	105.8
Lebanon	85.7	100.0	133.9	184.7
Malaysia	90.6	100.0	139.7	170.7
Mongolia	114.5	100.0	95.5	77.4
Myanmar	90.4	100.0	124.3	98.0
Nepal	111.1	100.0	101.8	116.1
Pakistan	97.7	100.0	103.5	112.1

TABLE 3.5 *Indices of food production per head by countries*

	1974–6	1979–81	1985–7	1990–2
Philippines	88.8	100.0	88.8	90.2
Saudi Arabia	214.3	100.0	188.6	353.6
Singapore	87.5	100.0	99.4	72.9
Sri Lanka	73.0	100.0	95.7	84.1
Syria	81.1	100.0	91.7	82.7
Thailand	93.2	100.0	104.6	108.7
Turkey	96.5	100.0	97.9	95.5
Viet Nam	90.9	100.0	111.9	123.0
Yemen Arab Republic	109.1	100.0	79.2	71.5
Yemen Democratic Republic	114.9	100.0	79.2	71.5
Europe				
Albania	92.6	100.0	94.5	76.1
Austria	94.9	100.0	109.0	105.5
Belgium-Luxemburg	96.2	100.0	104.9	120.2
Bulgaria	88.8	100.0	98.9	88.1
Czechoslovakia	97.4	100.0	116.7	115.2
Denmark	90.6	100.0	118.5	131.7
Finland	101.7	100.0	105.1	104.8
France	93.2	100.0	104.3	100.4
Germany DR	92.0	100.0	116.5	110.5
Germany FR	92.6	100.0	113.0	116.1
Greece	97.7	100.0	100.5	101.4
Hungary	90.6	100.0	109.9	106.3
Iceland	100.7	100.0	97.8	79.8
Ireland	96.8	100.0	111.2	121.5
Italy	90.9	100.0	100.3	97.9
Malta	79.2	100.0	107.3	100.3
Netherlands	94.9	100.0	109.8	110.1
Norway	93.5	100.0	103.7	108.7
Poland	107.9	100.0	107.0	106.6
Portugal	119.0	100.0	108.3	128.8
Romania	88.2	100.0	100.1	74.9
Spain	95.2	100.0	111.5	115.2
Sweden	97.1	100.0	102.1	92.0
Switzerland	91.5	100.0	104.8	104.7
UK	86.7	100.0	109.4	112.1
Yugoslavia	95.8	100.0	*	*
Oceania				
Australia	96.8	100.0	96.7	95.6
Fiji	76.3	100.0	88.5	95.3
New Zealand	95.6	100.0	108.5	101.8
Papua New Guinea	106.0	100.0	100.6	103.7
Solomon Islands	80.4	100.0	90.6	94.0
Tonga	103.4	100.0	90.6	83.8
Vanuatu	104.2	100.0	95.5	92.3
Former USSR	103.4	100.0	110.1	100.9

Source: See Table 3.2 above.

Barbados or Sri Lanka). Given the absence of a global food crisis, the ability of a country to *have* enough food really turns on its overall economic means rather than on its own food production, and it is in that direction we have to look to understand the crises in Sub-Saharan Africa.

The statistics of falling food output per head in Sub-Saharan Africa goes with reduction of general economic means. Sub-Saharan Africa experienced an average fall of GNP per head of 1.3 per cent per year during the decade 1980-90 (in contrast with the growth of GNP per head of 3.2 per cent per year in South Asia and 6.2 per cent annually in East Asia) (see *World Development Report* 1993: table A.2). While the declining trend in GNP per head in Sub-Saharan Africa is obviously influenced by its higher population growth rate (3.1 per cent *vis-à-vis* 2.2 and 1.6 per cent respectively in South Asia and East Asia), it is easily seen that the bigger part of the difference comes from the stagnation of production in Sub-Saharan Africa compared with economic growth in South and East Asia. Indeed, even if Sub-Saharan Africa had the same low population growth rate as East Asia, it would still barely escape having a falling GNP per head, and with the South Asian population growth rate it would continue to experience a sizeable decline. The interest in the population growth in Sub-Saharan Africa has to be supplemented very strongly by concern with ways of expanding production in this region.

In terms of the Condorcet–Malthus disputes, the use of appropriate incentives, and the utilization of science, technology, and methods of 'preservation', to which Condorcet had forcefully pointed, remain central in dealing with the predicament of Sub-Saharan Africa. It is not hard to identify the policy failures that have contributed to Sub-Saharan African production problems, and, more positively, the social and economic policies that can be fruitfully pursued to encourage the expansion of production of food and other commodities, and the creation of adequate economic means.[15] The policy instruments include such diverse items as making relative prices less unfavourable to food production, investing more on irrigation, expanding agricultural research, arranging more effective insurance facilities and credit arrangements, providing security of minimal entitlements, and so on. These policies are not unproblematic and there are several practical problems in implementation, even though the effectiveness of many of these policies has already been demonstrated in particular countries in Sub-Saharan Africa.[16]

The importance of preventing civil wars and military conflicts, which have had profoundly negative effects on many economies in this region, is also hard to overemphasize.[17] These have served as persistent barriers to the economic progress of Sub-Saharan Africa.

The opportunities of raising productivities in general and those in food production in particular are indeed quite extensive in Sub-Saharan Africa at this time. Without dismissing the issue of fast population growth, the production side of the problem requires appropriate emphasis, if only because, even with *zero* population growth, the progress of the Sub-Saharan economy would be very slow, unless its record of economic stagnation is reversed. In addition, of course, the problem of the high rate of growth of population in this region must also be considered. The factors that influence the birth rates will be taken up later (section 7).

6. APPROACHES TO PROBLEMS AND REMEDIES

When we broaden the analysis of the population problem from focusing narrowly on food, we have to consider the effects of demographic factors on the economy and society in general. Arguments have been presented suggesting that an increase in the labour force, even in the circumstances of poor countries, may contribute substantially to the expansion of production.[18] There is also a considerable literature on the effects of additional members of a family in raising family earnings (including through work performed by children). If these incomes are taken to be related to their productive contributions, then some prima-facie evidence would be found in the direction of presuming a positive impact of extra people on national production and social benefits.

These lines of reasoning are, however, made difficult by the divergence between private and social benefits, and also between private and social costs—for example, due to the presence of externalities and public goods (on this, see Birdsall 1988). There is also the issue of dichotomy between the interests of different members of the family (see Sen 1990). For example, the burden of child-bearing and rearing may fall mainly on women, whereas male household heads may be more influential in family decisions in many traditional societies (and may quite possibly give more weight to the additional outside incomes generated by child labour). The accounting of the suffering and social costs associated with child labour is another serious concern (see Dasgupta 1993*a*, 1993*b*).

It has also been suggested that, far from contributing much to national production and benefits, an expansion of population can have the effect of *reducing* the aggregate opportunities in the community and may lead to a negative impact on aggregate production and on well-being. Overcrowding can lead to over-use and diversion of land and other local resources enjoyed in common, and thus produce what Partha Dasgupta (1993*b*) has called 'the erosion of the local environmental resource-base'. Rapid population growth can also yield social problems with considerable negative effect on the well-being of the members of the community (for example, the impact of frequent pregnancies may be quite detrimental to women's health, aside from reducing their effective freedom to do various things which they may have reason to value).

If the population problem is not seen in terms of the Malthusian fear of population outrunning food supply, but in the light of the general impact of rapid population expansion on the well-being and freedom of women and men in the society, then there are indeed real problems to consider and investigate, which may not be as easily disposed of as the highly publicized anxiety about global food shortage. This way of seeing the issue is more in line with Condorcet's broader formulation of the problem, and there is little merit in taking the narrowly food-oriented Malthusian route.

The Malthus–Condorcet differences become even sharper when we consider the question of alleviating the problems that may be thus identified. The possible remedies may take different forms, such as fiscal policies to reduce the gap between social and private benefits, the creation of useful social institutions (varying from markets for some externalities and credit organizations to lend to the financially precarious), the political empowerment of those who have most to suffer from frequent births (including women), and so on.[19] What links the search for such remedies to the Condorcet side of the Malthus–Condorcet dispute are: (1) the absence of Malthusian fatalism and the willingness to look for solutions to difficult social problems (rather than accepting the inevitability of misery); (2) the procedure of rational assessment of the effectiveness of alternative social and economic policies; and (3) the ultimate reliance on voluntary, reasoned decisions (rather than compulsion) in bringing about suitable revisions of behaviour patterns.

7. FAMILY SIZE: VOLUNTARISM VERSUS COERCION

Nowhere is the contrast between Malthus and Condorcet as sharp as it is on the possibility of voluntary restriction of family size. As was discussed earlier, Malthus was convinced that, short of reducing people to hardship and misery, the growth of population could not be checked, and the so-called 'moral restraint' will not really work. As he put it, 'the perpetual tendency in the race of man to increase beyond the means of subsistence is one of the great general laws of animated nature which we can have no reason to expect will change' (1798: 198–9).

A vast demographic literature clearly indicates that this pessimism is unfounded. Birth rates do seem to go down with the availability of health facilities and contraceptives, with decline of death rates, with rise in economic prosperity, with female education, and so on.[20] To these can be added other—less immediate—connections, such as the impact of credit facilities, savings markets, and even, it has been suggested, the ready availability of water and fuel; they can reduce the private net benefits from procreation and bring them more in line with social concerns (Dasgupta 1993*b*).

In contrast, Malthusian pessimism about the prospect of voluntary family planning can serve as the basis for recommending compulsion in forcing people to cut the size of their families. While Malthus's 'positive checks' relate to the compulsion of misery and starvation in leading to smaller families, many neo-Malthusians would recommend having that compulsion through forceful state regulation of permissible size of the family, or of allowable number of child births.[21]

Among the developing countries, China has distinguished itself in going for compulsion in cutting down the growth rate of population, through such measures as the 'one-child policy' and through making basic social security and economic rights (such as housing) conditional on following the government's rules about the number of births. The Chinese birth rate has certainly fallen quite sharply; the last systematic calculation put it at around 21 per thousand—considerably lower than India's 30 per thousand, not to mention the average figure of 38 per thousand of poor countries other than India and China. This has been seen as fairly definitive evidence that compulsion works in reducing the number of births in developing countries, and that nothing else works so well.

These claims require more scrutiny. For example, China's birth rate is not lower than that of the state of Kerala in India. Within India, there are wide variations in birth rate, and these variations relate to diverse factors, including life-expectancy and mortality rates, education (especially female education), health facilities, and availability of means of family planning and medical help. Kerala stands out as a state that does well in all these respects, and the birth rate in Kerala has fallen sharply over the last few decades, from 44 per thousand in the 1950s to 20 per thousand in the late 1980s.

I should perhaps mention, to respond to a possible methodological concern, that, though Kerala is a state within federal India, it is by no means too small to be taken seriously in international comparisons. For example, its population of 29 million is rather larger than that of Canada. While Kerala is not one of the richer states in India, it has the highest life expectancy in India (more than 70 years—a little higher than China's) and the highest rate of literacy in general and female literacy in particular (higher than that in China as a whole and also, for the corresponding rural populations, higher than *every* province in China, particularly so in female literacy).[22] The Keralan birth rate of 20 per thousand is certainly no higher than the Chinese rate of 21 per thousand in the corresponding period, and this has been achieved not by compulsory birth control or the violation of any individual liberty to decide on these matters, but by the voluntary exercise of the family's right to family planning. Later, provisional statistics suggest that China's further fall in birth rate in very recent years (estimated to be moving towards 19 per thousand) has continued to be matched by Kerala's (calculated to be 18 per thousand in 1991). The respective fertility rates are similarly comparable.

Underlying the change in Kerala is the operation of economic and social incentives towards smaller families. As the death rate has fallen and family-planning opportunities have been combined with health care, and the desire of Keralan women—more educated as they are—to be less shackled by continuous child-rearing has become prominent, the birth rate has tumbled, in a way that would have astonished Malthus. What also appears to have played a part is a general perception that the lowering of the birth rate is a real need of a modern family—a conceptualization in which public education and enlightened discussion have been very effective. This last factor too relates closely to the social vision of Condorcet, and his particular emphasis on the role of open public discussion and the importance of female education.[23]

The appeal of compulsory birth control relates to pessimism about voluntary restriction of family size and the eagerness to institute state compulsion instead. That hiatus can lead to several deeply disturbing results. While China has ended up with a similar birth rate to Kerala's, one result of the compulsion has been a much higher level of mortality of female children. The traditional 'son preference' in China seems to have contributed to extreme reactions to compulsory birth-control measures, including possibly female infanticide and certainly a tendency to neglect the female child, yielding higher rates of female mortality among infants and children.

Gender bias is, of course, a widespread problem that applies to India as well as China.[24] The bias seems to be remarkably less in Kerala, possibly due to its long history of female education. In contrasting China with Kerala, it is instructive to note the general statistic that the ratio of females to males in the population, which is substantially higher than unity in Europe or North America, is still as low as 0.94 in China (rather like India's average 0.93). In contrast, the ratio in Kerala is well above unity, as in Europe and America.

One should not draw too many conclusions on the basis of this one contrast between Kerala and China, but since the Chinese success in cutting down birth rate is often cited in policy discussions, the comparison does have some relevance and force. Kerala's experience points, on one side, to the feasibility of cutting down birth rates, without the use of compulsion, even in a very poor economy, just as much as China has done with coercive methods. It also indicates that some of the accompanying evil effects of compelling people to do things they would not voluntarily choose to do can be avoided by taking a different route—one based on encouraging and creating conditions for reasoned decisions and enlightened rational behaviour. Condorcet had indeed seen something that Malthus had missed.

NOTES

1. Malthus (1798: 123). See also Wrigley and Souden (1986).
2. Kennedy 1993: 5–6. Kennedy has been particularly influential in directing attention to the social and organizational aspects of the population question, to supplement the usual focus on productive resources and the physical environment.
3. Malthus (1798: 123), emphasis added. In fact, even Malthus's translation of the passage of Condorcet is arguably somewhat defective to the extent that Malthus cites Condorcet as referring to the 'means of subsistence', whereas Condorcet (1795: 257) had spoken generally of 'means' (*moyens*), not necessarily of subsistence only. Condorcet's range of concerns in relating means to happiness was certainly much broader than Malthus's primarily food-centred view.
4. Condorcet 1795: 187–8. Elsewhere, Condorcet describes the changes in the production process through which the challenge would be met: 'A very small amount of ground will be able to produce a great quantity of supplies of greater utility or higher quality; more goods will be obtained from a smaller outlay; the manufacture of articles will be achieved with less wastage in raw materials and will make better use of them' (ibid.: 187).
5. Malthus saw the English 'poor laws' as contributing greatly to population growth, and having the effect of depressing 'the general condition of the poor' (1798: 96–7).
6. On this subject, see also Mitchell and Ingco (1993).
7. The figures are based on the data presented in United Nations (1993: table 4) and United Nations (1984: table 1). The comparison between 1974–6 and 1990–2 uses the 1979–81 average—common to the two time series—as the 'normalizer' (this yields an increase of more than 6.5%). It is possible that there may be some little variations in weighting between the two series, but the quantitative difference will be very small (certainly most unlikely to pull down—if that is what it does—the observed 6.5% figure below the 6% number suggested in the text).
8. There are many technical issues in these relationships, connected with technological and economic circumstances; I am not going into them here. On the plausibility and limits of characterizing the problem this way, see Meade (1955), Johansen (1972), and Dasgupta and Heal (1992).
9. See UNCTAD VIII (1992: table V-5). The period covered is between 1979–81 and 1988–90. The 'real prices' are 'based on price in current US$ divided by the UN index of export unit values of manufactured goods' (ibid.: 235).
10. The inclusion of externalities, including environmental effects, will tend to

change the relative prices of food and other goods. But the environmental worries apply at least as much to industrial production as they do to the increase in agricultural activities.

11. Interestingly enough, while Malthus had tended to concentrate rather wholeheartedly on the rhetoric of 'the proportion between the natural increase of population and food' in his *Essay on Population*, in his other works he was very acutely concerned with the role of effective demand and incentives in the operation of the economy. In fact, Keynes's (1936: 362–3) great praise of Malthus (citing Malthus's letter to David Ricardo, dated 7 July 1821, and the 'Preface' to his *Principles of Political Economy*) was related specifically to Malthus's analysis of 'effective demand'. That side of Malthusian economic analysis is also well illustrated by his deeply illuminating essay, *An Investigation of the Cause of the Present High Price of Provisions*, published in 1800, two years after his *Essay on Population*. I have discussed the far-reaching contribution of that analysis in Sen (1981: app. B). See also Wrigley and Souden (1986).

12. The Bengal famine of 1943 which killed between 2 and 3 million people began with a very sharp rise in the price of rice (despite the output being not much reduced), resulting from a war boom that added to the purchasing power of the urban poor but left rural labourers with little increase in income but much higher prices. The initial upward pressure on prices was compounded by other economic operations (including precautionary and speculative withdrawal of supplies, fed by panic). See Sen (1981: ch. 6, and apps. B and D).

13. On these and related matters on food entitlements and economic incentives, see Sen (1981) and Drèze and Sen (1989).

14. Indeed, as Peter Svedberg (1990) has shown, the extent of undernourishment in Sub-Saharan Africa, especially among children, is often substantially overestimated.

15. See, among others, Berry and Kates (1980), Eicher and Staatz (1984), World Bank (1984), Rose (1985), Commins, Lofchie, and Payne (1986), Eicher (1986), FAO (1986), Swaminathan (1986), Mellor, Delgado, and Blackie (1987), Lipton (1988), the essays in Drèze and Sen (1990), and Dasgupta (1993*a*).

16. Water sources may prove to be a barrier, in some regions, to the expansion of irrigation, and even to maintaining water supplies to the growing population. On the irrigation performance and prospects in Africa, see Moris and Thom (1985), and Moris (1987).

17. On these and related matters, see Sen (1992), and also Drèze and Sen (1989: sect. 13.6).

18. For a strongly argued and influential contribution in this direction, see Simon (1981).

19. See Dasgupta (1993*a*, 1993*b*) and the extensive literature cited there.
20. See e.g. Easterlin (1980). An interesting recent paper is Barro and Lee (1993).
21. An influential and widely read contribution advocating reduction of individual and family rights, so as to reduce the birth rate, can be found in Hardin (1993).
22. Drèze and Saran (1993).
23. In emphasizing the role of female education, Condorcet had argued that education 'can only really be taken proper advantage of when it has the support and encouragement of the mothers of the family' (1795: 193).
24. There is quite a vast literature on gender bias in China, India, and many other countries in Asia and Africa. A major new study of the nature of gender bias in India is Basu (1992).

REFERENCES

Barro, R. and Lee, J. W. (1993), 'Losers and Winners in Economic Growth', mimeo., Harvard University, presented to the World Bank Annual Conference on Development Economics, 3-4 May 1993, to be published in the proceedings.

Basu, A. (1992), *Culture, the Status of Women and Demographic Behaviour* (Oxford).

Berry, L., and Kates, R. (1980), *Making the Most of the Least* (New York).

Birdsall, N. (1988), 'Economic Approaches to Population Growth', in H. Chenery and T. N. Srinivasan (eds.), *Handbook of Development Economics*, i (Amsterdam), 477-542.

Commins, S., Lofchie, M., and Payne, R. (1986), *African Agrarian Crisis* (Boulder, Colo.).

Condorcet, M. J. A. N.(1795), *Sketch for a Historical Picture of the Progress of the Human Mind*, trans. J. Barraclough (London, 1955).

Dasgupta, P. (1993*a*), *An Inquiry into Well-being and Destitution* (Oxford).

Dasgupta, P. (1993*b*), 'The Population Problem', paper presented at the New Delhi conference of scientific academies on world population, 24–7 October 1993.

Dasgupta, P. and Heal, G. (1992), *Economic Theory and Exhaustible Resources* (Cambridge).

Drèze, J., and Saran, M. (1993), 'Primary Education and Economic Development in China and India: Overview and two Case Studies', STICERD, London School of Economics and Delhi School of Economics.

Drèze J. and Sen, A. (1989), *Hunger and Public Action* (Oxford).

Drèze, J. and Sen, A. (1990) (eds.), *Political Economy of Hunger* (Oxford).

Easterlin, R. E. (1980) (ed.), *Population and Economic Change in Developing*

Countries (Chicago).

Eicher, C. K. (1986), *Transforming African Agriculture* (San Francisco: Hunger Project).

Eicher, C. K. and Staatz, J. M. (1984) (eds.), *Agricultural Development in the Third World* (Baltimore).

FAO (1986), *African Agriculture: The Next 25 Years* (Rome).

Hardin, G. (1993), *Living within Limits* (New York).

Johansen, L. (1972), *Production Functions* (Amsterdam).

Kennedy, P. (1993), *Preparing for the Twenty-First Century* (New York).

Keynes, J. M. (1936), *The General Theory of Employment, Interest and Money* (London).

Lipton, M. (1988), 'The Place of Agricultural Research in the Development of Sub-Saharan Africa', *World Development*, 16.

Malthus, T. R., (1798), *Essay on the Principle of Population, As It Affects the Future Improvement of Society with Remarks on the Speculation of Mr. Godwin, M. Condorcet, and Other Writers*, Penguin Classics ed. (Harmondsworth, 1982).

Malthus, T. R. (1800), *An Investigation of the Cause of the Present High Price of Provisions* (London).

Malthus, T. R. (1830), *A Summary View of the Principle of Population* (Harmondsworth, 1982).

Meade, J. (1955), *Trade and Welfare* (Oxford).

Mellor, J. W., Delgado, C. L., and Blackie, C. L. (1987) (eds.), *Accelerating Food Production in Sub-Saharan Africa* (Baltimore).

Mitchell, D. O., and Ingco, M. D. (1993) 'The World Food Outlook', World Bank, Washington, DC, International Economics Department, November 1993.

Moris, J. (1987), 'Irrigation as a Privileged Solution in African Development', *Development Policy Review*, 5.

Moris, J. and Thom, D. (1985), *African Irrigation Overview* WMS Report 37 (Logan, Utah: Utah State University, Agriculture and Engineering)

Piel G. (1992) *Only One World* (New York).

Ram, N. (1990), 'An Independent Press and Anti-Hunger Strategies: The Indian Experience', in Drèze and Sen (1990).

Rose, T. (1985), *Crisis and Recovery in Sub-Saharan Africa* (Paris: OECD).

Sen, A. (1981), *Poverty and Famines: An Essay on Entitlement and Deprivation* (Oxford).

Sen, A. (1984), *Resources, Values and Development* (Oxford).

Sen, A. (1990), 'Gender and Cooperative Conflict', in I. Tinker (ed.), *Persistent Inequalities*, (New York).

Sen A. (1992), 'War and Famines', in W. Isard and C. H. Anderton (eds.), *Eco-*

nomics of Arms Reduction and Peace Process (Amsterdam).

Simon, J. (1981), *The Ultimate Resource* (Princeton, NJ).

Svedberg, P. (1990), 'Undernutrition in Sub-Saharan Africa: A Critical Assessment of the Evidence', in Drèze and Sen (1990).

Swaminathan, M. S. (1986), *Sustainable Nutritional Security for Africa* (San Francisco: Hunger Project).

UNCTAD VIII (1992), *Analytical Report by the UNCTAD Secretariat to the Conference* (New York: United Nations).

United Nations (1984), *FAO Monthly Bulletin of Statistics*, 7.

United Nations (1989), *UNCTAD Statistical Pocket Book* (New York: United Nations).

United Nations (1993), *FAO Quarterly Bulletin of Statistics*, 6.

World Bank (1984), *Towards Sustained Development in Sub-Saharan Africa: A Joint Program of Action* (Washington, DC).

World Development Report (1993) (Oxford).

World Bank (1989) *Price Prospects for Major Primary Commodities*, vol. ii, Food Products and Fertilizers (Washington, D.C.).

World Population News Service (1993), 'US. Urged to Adopt Policy on Population', in Popline: World Population News Service, 15, (Sept.–Oct. 1993).

Wrigley, E. A. and Souden, D. (1986), 'Introduction', in Wrigley and Souden (eds.) *The Works of Thomas Robert Malthus* (London).

4

An Ecologist View of the Malthusian Conflict

C. S. Holling[*]

I feel uncomfortable, as an ecologist, speaking to an interdisciplinary audience on the Malthusian conflict. That is because we so often use caricatures of other disciplines in order to give sharpness and definition to our own view, to our own understanding, and to our own desires for action. And, in recent years, the caricatures of ecology, as perceived by the public, have little relation to the scientific, biological, and systems traditions that have formed my own ecological research and related policy activities. Disciplinary knowledge is not static. It progresses in lurches of expanded understanding as theory and practice confront reality and as expanded debate with other disciplines widens comprehension. A useful stereotype at one moment can quickly lose its force as issues arise and understanding widens that can better define the human condition and responses to it. That is now the case for a wide range of present issues generated by the expanding and intensifying interaction between people, nature, and economic development.

Recently the Beijer Institute of the Royal Swedish Academy of Sciences sponsored a meeting between five ecologists and five economists in order to identify the differences in concepts, models, problem definitions, and policies between the two disciplines. In contrast to the stereotypes we each had of the other field, we discovered that there has been more convergence between the two disciplines than most of us had recognized. The ecologists learnt that the cornucopian arguments presented by Julian Simon (see e.g. Simon and Kahn 1984) did not at all represent the evolving body of theory and practice of economists at the leading edge of theory development, analysis and, practice. That is, growth in material possessions is not a given good, there are limits both to peoples' adaptive capabilities and to nature's resilience, and the pre-

* Arthur R. Marshall Jr. Professor in Ecological Sciences, Arthur R. Marshall Ecosystem Laboratory, Department of Zoology, University of Florida.

sent and expanding scale of human activities on the planet is stretching those limits. And the economists, in turn, discovered that ecologists active in the scholarship and practice of their science were only partly the bird-, bunny-, and bug-collecting stereotypes of nineteenth-century naturalists and not at all the no-growth environmentalist stereotype of the present social and political landscape.

Both the economists and the ecologists—or the ones at the meeting— saw a similar profile to the present problems that challenge traditional national and international problems:

- The problems are more and more frequently caused by local human influences on air, land, and oceans that slowly accumulate to trigger sudden abrupt changes that directly affect the health and innovative capacities of people, the productivity of renewable resources, and the vitality of societies.
- There is an increasing globalization of biophysical phenomena that couples with globalization of trade and with large-scale movements of people. That is, the spatial span of connections is intensifying so that the problems are now fundamentally intensifying across scales in space as well as in time. Local problems, and effective responses to them, can more and more often have their cause and their solutions half a world away.
- The problems are ones that emerge in several places and suddenly, rather than ones that emerge only locally and at a speed that is rapid enough to be noticed but slow enough to permit considered response. That is, they are fundamentally non-linear in causation and discontinuous in both their spatial structure and temporal behaviour. As a consequence, human responses that rely on waiting for a signal of change and then adapting to it will not work.
- The problems and the potential responses to them move both societies and natural systems into such novel and unfamiliar territory that aspects of the future are not only uncertain but are inherently unpredictable. That is, both the ecological and social components of these problems have an evolutionary character.

The purpose of this presentation is to review the advances in understanding and experience that led to this diagnosis from the perspective of ecology—particularly from those fields of ecology that deal with populations of organisms. I will do so by exploring three questions:

1. *Why are ecologists so gloomily Malthusian?* The answer, briefly, will be because there are demonstrable Malthusian forces in nature.

2. *Why, then, has the world not collapsed long ago?* The answer, briefly, will be that there are counteracting forces that give ecological systems the resilience and adaptability to deal with considerable change and that provide people with the capacity to innovate and create.

3. *Why, then, worry about the negative impacts of growth in human populations and activities?* The answer, briefly, will be that nature, people, and economies are suddenly now co-evolving on a planetary scale. Each is affecting the others in such novel ways and on such large scales that large surprises are being detected and posited that challenge traditional human modes of governance and management and that threaten to overwhelm the adaptive and innovative capabilities of people.

1. WHY ECOLOGISTS ARE GLOOMY MALTHUSIANS

For much of this century ecology has been more a biological tradition than an environmental one. Just as Darwin's theories of species-change were shaped by Malthus, so ecologists' theories have been rooted in the foundations of Darwinian natural selection and evolution. The environment has been seen principally as a fixed and exhaustible backdrop to the actions of and interactions among the biota in their confrontation with the physical environment. It was a demographer, Raymond Pearl (1927), who was one of the first to encapsulate the essence of this perception in a simple, and for that reason influential, representation of population growth. He showed that the populations of many organisms in the laboratory grew over time as an S-shaped curve to a plateau that was sustained if food resources were continually maintained. The logistic equation fitted many of those patterns of population growth, of yeast for example, or of *Drosophila*, with remarkably close fits, and the two parameters of the logistic began to be seen as reflecting two of the three axioms of ecology.

The parameter r, or the instantaneous rate of growth, represented the universal axiom that populations have the inherent propensity to grow exponentially, inevitably exceeding or reaching limits of the external environment. The parameter K, or the asymptote to which the logistic growth approached, in turn represented the axiom that the environment sets ultimate limits to growth.

Those two fundamental properties are viewed as being essentially self-evident and incontrovertible axioms by most biologists. In biologically based ecology, they appear in a number of forms. For example, one of the pioneers in ecology, Robert MacArthur, proposed that species of organisms can be designated as following one of two principal strategies as represented by the symbols r and K. The r-strategists are the pioneer species, the opportunists that can deal with the wide range of physical variability often found in recently disturbed habitats, but cannot compete well with other species. They are the 'weeds' of the biota. They have short lifetimes, small size, and small, widely dispersed, and abundant propagules. In contrast, the K-species are the conservative species with strong competitive abilities. They are able to out-compete the r-strategists, persist for long periods, and have long lifetimes, large size, and fewer but larger propagules. (See MacArthur and Wilson 1967; MacArthur 1972).

Populations of both groups encounter limits. The limits for the r-strategists are typically set by the growing dominance of the superior competitors, the K-strategists, and by the limited availability and accessibility of transient habitats. For the r-strategists, existence depends on being 'nimble-footed' in a world where disturbance generates spatial and temporal variety. The limits for the K-strategists are set by the more traditional Malthusian limits of space and resources—limits set by immutable laws of physics.

Even earlier than MacArthur, essentially the same two axioms provided the foundations for a plant ecologist, F. E. Clements (1916), to propose a theory of ecological succession in the 1920s that proposed that physical features of any area—temperature, precipitation, and geological substrate—allowed only one particular set of species to form a stable and persistent assemblage that he termed the climax. Such climaxes maintained themselves unless disturbed by some exogenous events. In the case of such a disturbance, different pioneer species, that is, r-strategists, could occupy the disturbed area because of their high dispersal properties and could survive because of their ability to withstand extremes of micro-climate. Once established, they could then moderate those extremes and the successional process could shift the species assemblage to those with the features of K-strategists or, as Clements termed them, of climax species. Again, the heart of the argument rested on the two axioms of exponential growth and ecological limitation.

MacArthur gave focus to the school of ecology called community ecology and Clements to the school that now would be called ecosystem ecology. At the same time as Clements was writing, still another tradition of ecology began to be developed. This was mathematical ecology, whose initial foundations derived from applied physics. It long remained detached from the empirical traditions of experimental and field-based ecology. This tradition was initiated by A. J. Lotka (1925) in a seminal book that continues to inform population ecology to this day. But it was one set of differential equations, the Lotka–Volterra equations, that captured most attention. These equations provide a simple conceptual model with two coupled differential equations that express the interactions between populations of predator and prey or between two competitors. Again, each has an exponential growth term and a population limitation term. Using the simplest expression for these terms, the equations generate population oscillations around an equilibrium point—formally, the type of oscillation termed a neutrally stable orbit. As I will note in the next section, however, the form of these expressions has subsequently been studied in the field and laboratory, leading to more realistic generalizations whose various alternatives generate a wide range of different population behaviours—asymptotic or stable-limit cycles, rarely, or, more commonly, local extinctions, multi-stable states, and chaotic behaviour.

Two Australians, Nicholson, an entomologist, and Bailey, an applied mathematician, were perhaps the first to give a sense of biological reality to mathematical ecology (Nicholson and Bailey 1935). Oddly, the growth and death terms of their equations were initially identical to those of the simple Lotka–Volterra equations, but they expressed theirs not as differential equations but as difference equations. The reason was a biological one—changes in insect-population numbers are more logically seen as a discontinuous process since, commonly, populations of one generation are completely replaced by the populations of the next generation. This introduced a simple lag which destabilized the behaviour so that the amplitude of population interactions between host and parasite or prey and predator increased until one went extinct followed by the other. If a lag is added to the simple Lotka–Volterra equations, the same thing happens.

But as simple, but biologically more realistic, assumptions of feedback controls were added, the oscillations stabilized to produce stable-limit cycles or to produce steady-state behaviour. A highly predictable, deterministic picture therefore gathered strength. It was an image of self-regulated populations in 'balance' with themselves and their environment, an image whose simplicity and ordering power provided the directions for a flood of research in field and laboratory. As a consequence, a revisionist view gathered force as nature began to be shown as much more interesting and variable than that static image suggests. Something of central importance was missing.

In summary, two of the fundamental axioms of ecological and evolutionary biology are that organisms are exuberantly over-productive, and that limits set by space, time, and energy are inevitably encountered. The foundations for all modern ecology and evolutionary biology rest in part upon the consequences of those two axioms. But, taken alone, they reinforce or produce a highly determined, static, and equilibrium-centred view which is not at all congruent with the way turbulent nature seems to behave. They seem to lead to a picture of an unforgiving nature that can be overwhelmed by change.

But there is a third element to this foundation that gradually has become seen as a third axiom. This concerns processes that generate variability and novelty. It was that third axiom that Darwin saw could transform a static view of a given fixed set of species on the planet to one of evolution of species. Exuberant over-production, interacting with limits, set the conditions for Darwinian 'survival of the fittest'. But it is the continual propagation of variation in species' attributes that provides the source for continued experimentation in an environment that itself changes. It is the result of the interaction of that third axiom of variability with the axioms of over-production and limits that provides the answer to my second question.

2. WHY HAS THE WORLD NOT COLLAPSED LONG AGO?

As those notably deterministic and equilibrium-centred views of the world matured, there was a continuing thread of opposition to them from those who saw relationships among organisms as being dominated by external variability, by diversity of the biota, and by variation in physical variables, and not by regulatory interactions among populations. If the latter view was one of 'Nature Balanced', this alternative view was 'Nature Anarchic'.

Thus, Gleason (1926) countered Clements's view of invariant succession moving inexorably to a sustained equilibrium climax, with a view of each individual species persisting as a consequence of its species-specific response to extant physical variability. In this view, the species present in an area did not maintain themselves because of the interactions among themselves but, rather, represented a random collection of species that happened to arrive at and be able to live in the same place under the same physical conditions.

Similarly, Andrewartha and Birch (1954) countered the deterministic view of their fellow Australians, Nicholson and Bailey, with a stochastic picture of local populations unstably increasing and collapsing in local places, at the whim of external physical variability. Persistence, in this view, was maintained by spatial heterogeneity. Collapsing local populations could be re-established by dispersal from other areas where populations were in an expanding phase. It was an anarchic view that saw all species essentially as *r*-strategists, a view reflecting experience with the climatic stress and physical variability of Australian ecosystems.

These two separate themes of biotic regulation versus physical variability began to converge in the early 1960s. One of the most influential steps in that convergence came from a famous experiment designed by Carl Huffaker (1958) at the University of California, Berkeley. This experiment demonstrated that, for strictly deterministic reasons, local populations can generate widely fluctuating changes in numbers that lead to local extinctions, and that spatial heterogeneity can indeed maintain persistence of populations in the large.

The experiment incorporated interactions among three entities—a predatory species of mite eating a plant-eating species of mite which, in turn, fed on oranges as its food resource. The oranges were laid out in a large room in a grid pattern so that local population interactions developed on each orange and a larger scale of interactions developed in the complete arena mediated by dispersal of mites between oranges. In present language, he essentially ran space-time simulations of population change using a living computer model in which the degree of dispersal was controlled between oranges by adding barriers to produce mazes of various complexities. The run time for each simulation, however, was measured in weeks, not seconds or minutes!

When dispersal was unimpeded in the full arena because of the absence of barriers, lagged oscillations of increasing amplitude consistently developed among predator and prey populations in the whole arena until the populations became extinct. The behaviour was similar to that generated by the simple Nicholson–Bailey equations. But, as barriers were added, those arena-wide oscillations became damped and self-sustaining. Local populations on each orange still went through a boom-and-bust pattern, but out of phase with each other, so that the whole system persisted.

Therefore, much as Andrewartha and Birch suggested, spatial heterogeneity induced by intermediate levels of dispersal turned local boom-and-bust populations into persisting interactions over a larger scale. But, much as Nicholson and Bailey argued, local predator–prey interactions could be unstable even under constant conditions in external physical variables. When variation in physical variables is added, a common pattern of shifting space-time dynamics emerges that generates great variation and diversity (of interactions, of species, and of genes) at small scales and yet persistence of assemblages at large scales. Spatial heterogeneity and the associated biotic diversity are an essential condition for the resilience a system needs in order to persist.

Since then, that image has been re-enforced and expanded in field and mathematical studies of population-patch dynamics by Bob Paine at the University of Washington and Simon Levin at Cornell, with John Steele of the Woods Hole Oceanographic Institution and Jonathan Roughgarden at Stanford adding critically important elements of large-scale variation induced by physical processes in marine systems (Levin and Paine 1974; Paine 1974; Roughgarden, Gaines, and Possingham 1988; Steele 1989).

My own early research contributed to this convergence between principles of population regulation and those of biotic and physical variation. Rather than simply assuming linearity in the way predators affected prey, as represented in the early models, my students and I engaged on a large-scale laboratory and field experimental programme in the late 1950s and 1960s involving everything from predatory insects (praying mantis—see Holling 1966), to fish (danio and salmon—see Holling 1959; Dill 1974; Dill and Fraser 1984), to predatory mammals (shrews and lions—Elliot, McTaggart Cowan, and Holling 1977). The goal was to develop a model, or a family of models, that could be proved to be both realistic and general and, hopefully, could be simplified. The resulting models were applied in a series of papers to explain an even wider range of situations, from predation by mites, to birds, to wolves.

Parenthetically, Ric Charnov (1976), during a post-doctoral period at my laboratory, combined the data and time-budgeting structure of this predation model with economic theories of marginal utility optimization, and by so doing established the foundation for an entirely new field termed optimal foraging theory. Other economists, mathematicians, biologists, and ecologists like Colin Clark (1991), Larry Dill (1987), John Krebs (Krebs and Davies 1978), and Marc Mangel (Mangel and Clark 1988) have evolved that theory into a distinct school of ecology—that is, behavioural ecology. Although this school is only tangentially related to the topic of this paper, it is worth highlighting because it is one of the first examples of a very fruitful union between ecological and economic models and concepts.

The basic predation experiments and models showed that all cases of predation fell into four functionally different classes of attack response, and two classes of competition response. Each of these classes had different consequences for the stability and instability of interactions, a point I shall return to in a moment, and each could be represented by a particular limiting condition of a simple, realistic, and general predation model with four easily estimated parameters.

Such a process model provides a well-tested module of one critical population process that can be combined with similar modules of reproduction, parasitism, competition, foraging, dispersal, and responses to physical variables to provide a set of building blocks for the development of large-scale models of multi-species population interactions or of ecosystem dynamics. Such models not only provide insight into the way natural ecosystems and populations function; they have also contributed to the development of resource and ecological management policies and to insights into the way human institutions interact with natural systems.

Those experiences have formed the foundations for what is known as adaptive environmental and resource management, an approach created by Carl Walters (1986) at the University of British Columbia as much as by me (Holling 1978). A series of examples, tests, and policies have been collaboratively developed around the world in the past fifteen years, organized initially because of the innovative environments provided by two organizations—the Institute of Resource Ecology at the University of British Columbia, and the International Institute of Applied Systems

Analysis in Laxenburg, Austria. The early, foundation studies emerged from joint studies with a number of colleagues—of forest–pest–fire dynamics and management with Gordon Baskerville (1988) and Bill Clark (Clark, Jones, and Holling 1979; Holling 1980), of aquatic systems and fisheries management with Carl Walters (1981), of savanna ecosystem dynamics and grazing management with Brian Walker (Walker *et al.* 1981), and of wetland function and water management with Lance Gunderson (Holling, Gunderson, and Walters 1994).

The answer to the question 'Why has the world not collapsed?' lies in the conclusions that can be drawn from two aspects of this body of research and application. One deals with the stability properties of natural populations and systems. The other deals with the way people and human institutions adapt and learn.

I shall first deal with the stability issue. The conclusions of these streams of work are unambiguous. Any of the realistic representations of the key processes show the existence of thresholds, limits, and other nonlinearities. As a consequence, each process can show population-density regions of negative feedback control, other regions of positive feedback destabilization, and still other regions of neutral influence. Once the models incorporate three or more population variables or species, together with realistic representations of the key processes, a very wide range of complex population behaviours is produced. Even in the simplest models, multiple stable states are the rule, not the exception, and behaviour can range from extinctions, to stable limit cycles, to boom-and-bust flips between stability regions, and even to chaotic behaviour, as Robert May (1974) at Princeton University first demonstrated.

All the examples of modelling terrestrial ecosystems noted above show that the interplay of natural processes and populations generate a nested set of quasi-limit cycle and boom-and-bust fluctuations organized by clusters of variables of distinctly different speeds (from days to centuries) that form changing multiple equilibrial states. The control of those behaviours lies in the interactions among a small set of biotic, abiotic, and geophysical variables, each set operating over a different range of spatial and temporal scales. Overall, those scales range from centimetres and days to hundreds of kilometres and centuries. As Nathan Keyfitz (1993) pointed out recently in a lecture to the US National Academy of Sciences, the range covered by such time scales is several orders of magnitude greater than the range of scales emphasized by economists and politicians. It is no wonder, therefore, that there is disagreement between many economists and ecologists over the effects of human activities and of human population growth on linked social and ecological systems.

Pelagic marine ecosystems, however, are different from the terrestrial ones emphasized to this point. Because of the physical attributes of water, as distinct from those of air, there is little capability in marine ecosystems for the biota to control variability. Organisms are forced to adapt to the existing physical variability at all scales, from local wave action to coastal currents and ocean-wide gyres. That is why some marine ecologists, such as Jonathan Roughgarden at Stanford, see ecology as less of a biological and more of an earth-science discipline (Roughgarden, Pennington, and Alexander 1994).

But for terrestrial systems, the geophysical controls only dominate at scales larger than kilometres. At scales smaller than this, biotic processes, interacting with abiotic ones, can control structure and variability. These are also the scale ranges where human land-use transformations occur, so that the arena where plant and animal controlling interactions unfold is the same arena where human activities and population interact with the landscape. That is why human population growth is so inexorably interconnected with terrestrial ecosystem resilience.

The controls determined by each set of biotic structuring processes within terrestrial ecosystems are remarkably robust and the resulting behaviours are remarkably resilient.

By 'robust' I mean that there is so much functional diversity and spatial heterogeneity in the keystone structuring set of processes that their regulatory role retains its integrity in the face of great changes in populations of the keystone set species or in values of the keystone physical variables.

By 'resilient' I mean that the stability domains that define the type of system (for example, forest, savanna, grassland, or shrub steppe) are so large that external disturbances have to be extreme and/or persistent before the system flips irreversibly into another state. Except under extreme climatic conditions, Mother Nature is not basically in a state of delicate balance. If she was, the world would have collapsed long ago.

The metaphors of Nature Balanced and Nature Anarchic, therefore, have to be expanded to include Nature Resilient. So long as we only perceive the axioms of exponential growth and environmental/ecological limits, then the Malthusian determinism of Nature Balanced seems inexorable and near equilibrium behaviour the only behaviour of interest. In contrast, when we only perceive external physical variability and a passively adapting biota, then Nature Anarchic is the logical image and spatial heterogeneity emerges as the critical ingredient for persistence in a world of locally unstable equilibria.

When, however, we perceive a structuring and controlling role for keystone clusters of the biota at lower-scale ranges, for zootic and abiotic processes like insect outbreaks, large-ungulate grazing, and storm and fires at intermediate-scale ranges, and for geophysical processes at large-scale ranges, then the image of Nature Resilient emerges. Such an image incorporates the principles of population regulation and of physical variation that are contained in the two other metaphors, but adds principles of biotically induced variation.

In this view, behaviours near equilibrium and all the traditional mathematical tools for local stability analysis are irrelevant. Populations assume trajectories that take them far from the structuring equilibrium attractors and repellers. The critical focus then becomes conditions at the boundaries of stability domains, the size of those domains, and the forces that maintain those domains. That was what I emphasized in a review paper entitled 'Resilience and Stability of Ecological Systems' (Holling 1973) as an antidote to the narrow view of fixed, equilibrium behaviour and of resistance of populations to local perturbation. Those narrow, essentially static notions have provided the foundations for the now-discredited goals of maximum sustained yields of fish populations or of fixed carrying capacity for terrestrial animal populations. The success of achieving such goals squeezes out variability and resilience is lost.

The three axioms: over-production of populations, environmental limits, and physical and biotic variation, therefore, become as fundamental for ecology as they are for evolutionary biology and natural selection.

So part of the answer to the question 'Why has the world not collapsed?' is that natural ecological systems have the resilience to experience wide change and still maintain the integrity of their functions. The other part of the answer lies in human behaviour and creativity. Change and extreme transformations have been part of humanity's evolutionary history. People's adaptive capabilities have made it possible not only to persist passively, but to create and innovate when limits are reached. At its extreme, these attributes underlie many economists' pre-

sumptions that peoples have an unlimited capacity to substitute for
scarce materials and to develop successful remedial policies incremen-
tally once the need is apparent.

But the resilience of natural systems, while great, is not unlimited. And
the effective exercise of peoples' adaptive and creative abilities relies on
specific contexts set by the environment. Are those contexts changing as
human populations and activities expand? It is those limitations rather
than the Malthusian ones of resource limitations, that make the third
question relevant.

3. WHY WORRY ABOUT THE NEGATIVE IMPACTS OF GROWTH IN HUMAN POPULATIONS AND ACTIVITIES?

Now I can return to the studies of ecological management and policy
design mentioned earlier and explore the second set of conclusions concerning the way people adapt and learn. Those cases involved a number
of different examples of forest development, of fisheries exploitation, of
semi-arid grazing systems, and of disease management in crops and people. Since the 1970s we have greatly expanded and deepened the case
studies and tests—again with collaborators. The cases cover forest management in New Brunswick (Baskerville 1994); water management for
agriculture, cities, and ecosystems of the greater Everglades system in
Florida (Light, Gunderson and Holling 1994); estuarine management in
the Chesapeake Bay (Costanza and Greer 1994); salmon and power in
the Columbia River (Lee 1994); water quality, fisheries, and development in the Great Lakes (Francis and Regier 1994); and the same issues
in the Baltic Sea (Jansson and Velner 1994). Those examples have become the focus for a new book co-edited with my colleagues Lance
Gunderson and Steve Light that combines an analysis of ecosystem
structure and dynamics with one of human organizational structure and
dynamics (Gunderson, Holling, and Light 1994).

One of the puzzles we encountered early in a comparison of these case
studies of ecosystem management was the apparently inevitable emergence of a pathology of regional development, at least in its early phases.
Briefly, as the consequence of initially successful exploitation and management, the ecosystem and environment became more vulnerable to
surprise and crisis. Resilience decreased because of loss of species and

of genetic diversity and the increase in spatial homogeneity of critical variables. Those changes took place gradually over decades. During that period, at a somewhat faster pace, the management institutions gradually became more efficient, but more rigid and less responsive both to the resource dynamic and to the public. Even more rapidly, the citizenry and industry became more dependent and with fewer options for self-reliance. That seems to describe an ultimate and perhaps incurable pathology—the ecosystem becoming less resilient, the management more rigid, and the society more dependent.

Two of the initial case studies—forest management in New Brunswick and fisheries issues in the Pacific Northwest—subsequently provided the insights needed to clarify the pathology and expose the way out of it. In those two examples we initially diagnosed the pathology up to the mid-1970s as follows (Holling 1986):

- Successful suppression of spruce budworm populations during the 1950s and 1960s in eastern Canada, using insecticide, an innovation from the Second World War, certainly preserved the pulp and paper industry in the short-term by partially reducing defoliation by the insect so that tree mortality was delayed. This encouraged expansion of pulp mills but left the forest, and hence the economy, more vulnerable to an outbreak that would cause more intense and more extensive tree mortality than had ever been experienced. That is, the short-term success of spraying led to moderate levels of infestation that spread and persisted over larger areas, demanding ever-more vigilance and control, at a time of growing public opposition to the use of insecticide, and increasing costs in its use.

- Effective protection and enhancement of salmon spawning through use of fish hatcheries on the west coast of North America quickly led to more predictable and larger catches by both sport and commercial fishermen. That triggered increasing fishing and investment pressure in both sectors, pressure that caused more and more of the less productive natural stocks to become locally extinct along the coast of North America. That left the fishing industry precariously dependent on a few artificially enhanced stocks, whose productivity began declining in a system where larger-scale physical oceanic changes began to contribute to unexpected impacts on distribution and abundance of fish.

In both those cases, however, in the 1980s we began to realize that the phase of a growing pathology was transient and was broken by a spasmodic readjustment, an adaptive lurch of learning that created new opportunity by beginning to revive spatial and species diversity at a regional scale. It is that creation of something fundamentally novel at a different scale that gives an evolutionary character to development of a region that might make sustainable development an achievable reality rather than an oxymoron. So even Nature Resilient is an incomplete representation. Nature Evolving adds the final ingredient of novelty and transformation in the relationships between nature and humanity.

That would end this exposition on a happy note, if it were not for continuing human population growth in the developing world and the declining or stable and ageing populations in the developed world. Increasing human populations in the South, and the planetary expansion of their influence, combined with exploitative management in both North and South, reduces functional diversity and increases spatial homogeneity not only in regions but on the whole planet. Functional diversity of the structuring processes and spatial heterogeneity are the two most critical determinants of ecological robustness and resilience, the attributes that provide the reserve of ecological services and of time that have allowed people to adapt and learn in the past. And now those critical attributes are being compromised at the level of the planet. For the first time, humanity and nature are transforming each other on the whole planet, are beginning a co-evolutionary experiment on a planetary scale.

Although natural systems are resilient, they are not infinitely so, and in exploitative situations that resilience shrinks. Therein lies the irony of ecological resilience—on the one hand, it provides the buffer for incomplete knowledge, therefore allowing experiment and recovery; but on the other, it also exacts few penalties fast enough on the greedy or the stupid. Ironically, is the reserve and insurance that the resilience of nature has given us going to be the very attribute that blocks people's abilities to perceive and adapt to abrupt change on planetary scales?

Thus the diagnosis described at the beginning profiles a context that requires us to worry very much about expanding human populations and of the effectiveness of human adaptive capabilities:

- The problems are more and more frequently caused by local human influences on air, land, and oceans that slowly accumulate to trigger sudden abrupt changes that directly affect the health and innovative capacities of people, the productivity of renewable resources and the vitality of societies.

- There is an increasing globalization of biophysical phenomena that couples with globalization of trade and with large-scale movements of people. That is, the spatial span of connections are intensifying so that the problems are now fundamentally intensifying across scales in space as well as in time. Local problems, and effective responses to them, can more and more often have their cause half a world away.
- The problems are ones that emerge in several places and suddenly, rather than ones that emerge only locally and at a speed that is rapid enough to be noticed but slow enough to permit considered response. That is, they are fundamentally non-linear in causation and discontinuous in both their spatial structure and temporal behaviour. As a consequence, human responses that rely on waiting for a signal of change and then adapting to it will not work.
- The problems and the potential responses to them move both societies and natural systems into such novel and unfamiliar territory that aspects of the future are not only uncertain but are inherently unpredictable. That is, both the ecological and social components of these problems have a co-evolutionary character at regional and planetary scales.

There have been other times when humanity has suddenly developed a connection with natural systems without a long period of slow co-evolutionary adaptation. Historians like William McNeill (1979) and A. W. Crosby (1986) spell out the disastrous consequences. One example occurred when Europeans discovered and occupied North America. The transfer of such city-nurtured diseases as smallpox, measles, and mumps to virgin populations with no acquired immunity literally decimated populations in North and South America. But, as well, the new invaders carried an assemblage of co-adapted organisms with them that had long evolved a relationship with humans practising sedentary agriculture in the Middle East and Europe. This aggressive assemblage of biota overwhelmed native fauna in the temperate regions of the world from Australia to North America, in an act of what Crosby colourfully called 'ecological imperialism'. The sudden connection of flora and fauna that had separately co-evolved under different circumstances causes a fundamental transformation in the least resilient of those assemblages.

An equally dramatic example affecting the whole western hemisphere occurred 11,000 years ago, with the migration of people from Siberia into North America near the end of the Pleistocene. Paul Martin (1984) of the University of Arizona has accumulated evidence to suggest that this human invasion accounts for the sudden extinction of a large proportion of the megafauna at the end of the Pleistocene. In much less than 1,000 years, 73 per cent of the genera of herbivores weighing over 44 kg became extinct. For similar reasons, South America and Australia were also affected, with, respectively, 80 per cent and 86 per cent of the mega-herbivores becoming extinct. In contrast, Africa and tropical Asia did not experience massive extinctions of the megafauna at that time, presumably because of the gradual co-evolution of people and animals that occurred over long histories of habitation in those more climatically stable regions.

Simulations using a so-called 'blitzkrieg' model suggest it would have been possible to achieve such extinctions in North and South America through hunting alone, as new human invaders swept in a wave-front, typical of invading species, from Alaska to Patagonia at an average rate of 160 km per decade. The timing of those extinctions is coincident with the appearance of peoples of the Clovis hunting cultures in the Americas. Since all species of the very large grazing and browsing herbivores above 1,000 kg (mammoths, gomphotheres, ground sloths, and mastodons) became extinct, the structuring influence of those very large herbivores on vegetation would have been lost. Human use of fire could then have led rapidly to vegetation cover of one type over extensive areas, thereby accelerating declines of the megafauna. It is again an example of the consequence of sudden engagement of relationships between humans and nature when previously they had been disengaged.

That is the source of worry now, because for the first time it can be demonstrated that human activities are exerting a new and sustained influence on the planet as a whole. The increases of the greenhouse gases, CO_2 and methane, and the thinning of the protective ozone layer in the stratosphere are unambiguously caused by intensifying human activity in the last several decades. The fact of those changes is incontrovertible. The specifics of the consequences are uncertain, but clearly include increases in climatic variability, possibilities of climate warming, increasing soil aridity in large areas, increasing UV radiation damaging plankton productivity in the oceans, and a host of interconnected phenomena.

Those particular examples are mediated by planetary atmospheric processes. But there are other examples, less well known, of equally large-scale and novel transformations taking place mediated by different planetary connections. I shall give three examples—one where the planetary flows are monetary, one where they are migrating animals, and one where they are movements of people.

Example 1: The Costs of Disasters

The *Wall Street Journal Europe*, Tuesday, 31 August 1993, carried an article describing how the dramatic increase in costs of natural disasters had prompted Swiss Reinsurance Company to found a new property-catastrophe reinsurer. According to Swiss Reinsurance Company data, between 1970 and 1986 global damages were below $5 billion on an annual average in 1992 dollars. In 1986 damages suddenly started to swell, reaching an unprecedented peak of $22.5 billion in 1992. Reinsurers, hit hard by burgeoning claims, started withdrawing from the market, leaving a massive need for catastrophe insurance coverage. In the United States alone, insurance demand of about $25 billion is estimated currently to face a supply of $3 billion to $4 billion. It now seems that not only insurance companies need reinsurers but reinsurers themselves need the buffer of still larger reinsurers.

Company officials believe the trend will continue and could accelerate. The causes are ascribed to climate change and increasing population pressure. The first effect is simply a consequence of more people moving to more vulnerable places. The second effect is the loss of resilience that comes from human constructions—such as dams, dikes, and drainage canals—that decrease the frequency of disaster, but do so by increasing the costs when the disaster inevitably occurs. The third effect is the established increase in the variability of climate. But if increases in greenhouse gases further destabilize climate, then positive feedbacks will have been developed between human population increase, increase in use of fossil fuels, and increase in deforestation on the one hand, and decreased regional-to-global resilience on the other, to precipitate a spiralling explosion in the costs of disasters.

Example 2: How the Forest Industry in Boreal Regions can be Affected
　　by Human Population Growth in Neo-Tropical countries

In 1987 I received an E-mail message from Gordon Baskerville, then of
the University of New Brunswick in Canada, remarking, in passing, that
insect outbreaks by a defoliating insect, the spruce budworm, were occurring under conditions and on tree species that previously had never
been observed. Now Baskerville's knowledge of boreal forest ecology
and of forestry policy and management, in my mind, is quite simply the
best in the world. And the spruce budworm had been the target of one of
the largest insect-pest spraying programmes in the world since the early
1950s, in order to allow for sustained exploitation of balsam trees for the
pulp and paper industry.

Thus it seemed possible that this surprise might be one example of the
class of surprising observations that suddenly expose the fact that slow
changes had been occurring over long periods over large parts of the
planet. There could be a number of reasons—climate-change effects, or
changes in spatial diversity within the region because of harvesting and
insect-control practices. But the one I chose to explore, more as an exercise in methodology than anything else, was the impact of possible
changes that could have occurred in the suite of thirty-five species of
insectivorous birds that we know are one of the controllers of budworm
populations at low densities in stands of younger or more scattered trees.
They are part of one of the robust processes mentioned earlier, ones that
are responsible for some of the space-time structures of the eastern
Boreal forest ecosystem.

A large proportion of the bird species migrate to spend their winters in
Central America and parts of South America. And people like John Terborgh (1989) had been accumulating evidence that their populations
were declining because of habitat fragmentation in temperate regions and
because of landscape transformations caused by human pressures in neo-
tropical countries.

I posed the question: 'How much change would there have to be in
insectivorous bird populations to cause a sudden, qualitative change in
insect outbreak patterns?' I used a model of budworm-forest dynamics
that had been part of the set of ecosystems studies mentioned earlier, and
through simulation experiments concluded that populations of these
birds would have to be reduced by about two-thirds before the systems

would flip into a different pattern of behaviour. Since such a large decline would probably be accompanied by extinction of several species, and such had not been observed, I concluded that it was unlikely that the surprising observations were directly connected to changes induced in bird populations (Holling 1988).

But the conclusion was wrong. Just one month ago I heard of a study by Sidney Gauthreaux (1992) comparing radar images of flights of migratory birds across the Gulf of Mexico taken in 1965–7 with those taken between 1987 and 1989. The frequency of trans-Gulf flights had declined by almost 50 per cent over that roughly twenty-year period. Moreover, it is now clear that at least one species of warbler, the Bachman's, is now extinct and another, Kirtland's Warbler, is perilously close to extinction.

This magnitude of decline is sufficient to make it credible that declines of bird populations that over-winter in the neo-tropics are contributing to changes in the structuring of the boreal forests of eastern Canada. It is indeed possible that the state of the pulp-and-paper industry in that region is being affected by human population growth and land-use pressures in neo-tropical countries.

Example 3: The 'Birthing' of New Human Diseases

In a landmark conference sponsored by the Rockefeller University and the US National Institute of Health in May 1989, growing evidence was presented that new infectious diseases are initially caused not by changes in pathogens but, more fundamentally, by changes in environments. Since then the issue has been explored in the News and Comment Section of *Science* (19 January 1990; 6 August 1993), in a book reviewing the conference (Morse 1993), and even in news magazines (*Newsweek*, 22 March 1993).

As reported in *Science* (Research News Section, 1993), Stephen Morse summarizes the issue as one in which 'new' viruses that have emerged in humans are in fact old viruses that have been around in other hosts or have existed in benign forms in humans for centuries. Examples include AIDS, Ebola, Marburg and yellow-fever viruses that probably occurred first in monkeys, Rift Valley fever in cattle, sheep- and mosquitoes and, the Hantaan virus, that recently caused its first ever recorded human deaths in Nevada, in rodents.

In one sense that is nothing particularly new. The historian William McNeill (1979) had much earlier brilliantly reconstructed the history of the emergence of diseases and their impacts on society over human history from pre-agricultural periods to the present. It is a story of emergence of new epidemics because of changing densities of people, changing land use, and movements of people.

But two new elements are now being added. One is that human movement is now both rapid, large, and planetary in scale. Another is that growing populations cause intensifying transformations of landscapes that reduce ecosystem resilience in more and more localities and regions.

The three axioms of population over-production, limits, and variation are the foundations of a new theory of evolutionary dynamics of diseases that makes the connection between new disease, population growth, and environmental change direct. Viruses, for example, are as much under the influence of natural selection as any other organism. The types that persist are simply the ones that leave the most progeny—that exploit their hosts most efficiently. On the one hand, if a microbe reproduces too aggressively in a host, it might kill its host and itself without being passed along to a new host. On the other hand, if it reproduces too slowly, insufficient numbers will be produced to allow for transmission to other hosts. Hence, virulence depends on a set of relationships within the host in relation to different modes of transmission and of microbial survival outside the host. Change any of those relationships and the virulence can change.

Paul Ewald (1994) of Amherst College argues, for example, that HIV may have infected people for centuries, causing little illness, before it suddenly started causing AIDS. In isolated African populations with modest levels of promiscuity, HIV would have no advantage in reproducing aggressively within the body. That would simply speed up its own extinction. But starting in the 1960s, war, tourism, and trucking connected once-isolated villages. Drought, foreign agricultural aid, and industrialization led to changed land-use patterns and rapid urbanization that shattered social structures and destroyed traditional patterns of movement that had constrained sexual behaviour. Those are just the conditions needed for natural selection to evolve a more virulent strain. Once initiated, the rapid global movements of people and freer life-styles in the developed countries created a global pandemic.

The causative forces behind those trends are not moderating. They are intensifying and expanding as populations in the developing world grow and as greater land use and social transformations occur in both north and south.

Why worry about population growth? Because new connections are rapidly forming and intensifying between people and nature on a planetary scale. We are now in the early destabilized phase of a process of co-evolution between humans and natural systems on a planetary scale. The future is not just uncertain; it is inherently unpredictable.

REFERENCES

Andrewartha, H. G., and Birch, L. C. (1954), *The Distribution and Abundance of Animals* (Chicago).

Baskerville, G. L. (1988), 'Redevelopment of a Degrading Forest System', *Ambio*, 17: 314–22.

Baskerville, G. L. (1994), 'The Forestry Problem', in Gunderson, Holling, and Light (1994).

Charnov, E. L. (1976), 'Optimal Foraging: The Marginal Value Theorem', *Theoretical Population Biology*, 9: 129–36.

Clark, C. W. (1991), 'Modeling Behavioral Adaptations', *Behavioral and Brain Sciences*, 14: 85–117.

Clark, W. C., Jones, D. D., and Holling, C. S. (1979), 'Lessons for Ecological Policy Design: A Case Study of Ecosystem Management', *Ecological Modelling*, 7: 1–53.

Clements, F. E. (1916), 'Plant Succession: An Analysis of the Development of Vegetation', *Carnegie Institute of Washington Publication*, 242: 1–512.

Costanza, R., and Greer, J. (1994), 'The Chesapeake Bay and its Watershed: A Model for Sustainable Ecosystem Management?' in Gunderson, Holling, and Light (1994).

Crosby, A. W. (1986), *Ecological Imperialism* (Cambridge).

Dill, L. M. (1974), 'The Escape Response of the Zebra Danio (Brachydanio Rerio), I. The Stimulus for Escape', *Animal Behaviour*, 22: 711–22.

Dill, L. M. (1987), 'Animal Decision Making and its Ecological Consequences: The Future of Aquatic Ecology and Behaviour', *Canadian Journal of Zoology*, 65: 803–11.

Dill, L. M., and Fraser, A. H. G. (1984), 'Risk of Predation and the Feeding Behavior of Juvenile Coho Salmo (Oncorhynchus kisutch)', *Behavioural Ecology and Sociobiology*, 16: 65–71.

Elliot, J. P., McTaggart Cowan, I., and Holling, C. S. (1977), 'Prey Capture by the African Lion', *Canadian Journal of Zoology*, 55: 1811–28.

Ewald, P. (1994), *The Evolution of Infectious Disease* (Oxford).

Francis, G. R., and Regier, H. A. (1994), 'Barriers and Bridges to the Restoration of the Great Lakes Basin Ecosystem', in Gunderson, Holling and Light (1994).

Gauthreaux, S. A., jun. (1992), 'The Use of Weather Radar to Monitor Long-Term Patterns of Trans-Gulf Migration in Spring', in J. M. Hagen III and D. W. Johnston, *Ecology and Conservation of Neotropical Migrant Landbirds* (Washington, DC).

Gleason, H. A. (1926), 'The Individualistic Concept of the Plant Association', *Bulletin of the Torrey Botanical Club*, 53: 7–26.

Gunderson, L. H., Holling, C. S., and Light, S. S. (1994) (eds.), *Barriers and Bridges to the Renewal of Ecosystems and Institutions* (New York).

Holling, C. S. (1959), 'The Components of Predation as Revealed by a Study of Small Mammal Predation on the European Pine Sawfly', *Canadian Entomology*, 91: 293–320.

Holling, C. S. (1966), 'The Functional Response of Invertebrate Predators to Prey Density', *Memoirs of the Entomological Society of Canada*, 48: 1–86.

Holling, C. S. (1973), 'Resilience and Stability of Ecological Systems', *Annual Review of Ecology and Systematics*, 4: 1–23.

Holling, C. S. (1978), *Adaptive Environmental Assessment and Management* (London).

Holling, C. S. (1980), 'Forest Insect, Forest Fires and Resilience', in H. Mooney *et al.* (eds.), USDA Forest Service General Technical Report, 445–64.

Holling, C. S. (1986), 'The Resilience of Ecosystems: Local Surprise and Global Change', in W. C. Clark and R. E. Munn (eds.) *Sustainable Development of the Biosphere* (Cambridge), 292–317.

Holling, C. S. (1988), 'Temperate Forest Insect Outbreaks, Tropical Deforestation and Migratory Birds', *Memoirs of the Entomological Society of Canada*, 146: 21–32.

Holling, C. S., Gunderson, L. H., and Walters, C. J. (1994), 'The Structure and Dynamics of the Everglades System', in S. Davis and J. Ogden (eds.), *The Everglades: Spatial and Temporal Guidelines for Ecosystem Restoration*, (forthcoming).

Huffaker, C. B. (1958), 'Experimental Studies on Predation: Dispersion Factors and Predator–Prey Oscillations', *Hilgardia*, 27: 343–83.

Jansson, B.-O., and Velner, H. (1994), 'The Baltic—The Sea of Surprises', in Gunderson, Holling, and Light (1994).

Keyfitz, N. (1993), 'Are there Ecological Limits to Population?', *Proceedings National Academy of Sciences*, Washington DC., 90: 6895–99.

Krebs, J. R., and Davies, N. B. (1978), *Behavioral Ecology: An Evolutionary*

Approach (Oxford).

Lee, K. N. (1994), 'Deliberately Seeking Sustainability in the Columbia River Basin', in Gunderson, Holling, and Light (1994).

Levin, S. A., and Paine, R. T. (1974), 'Disturbance, Patch Formation and Community Structure', *Proceedings of the National Academy of Sciences*, 71: 2744–7.

Light, S. S., Gunderson, L. H., and Holling, C. S. (1994), 'The Everglades: Evolution of Management in a Turbulent Ecosystem', in Gunderson, Holling, and Light (1994).

Lotka, A. J. (1925), *Elements of Mathematical Biology* (New York).

MacArthur, R. H. (1972), *Geographical Ecology* (Princeton, NJ.).

MacArthur, R. H., and Wilson, E. O. (1967), *The Theory of Island Biogeography* (Princeton).

McNeill, W. H. (1979), *The Human Condition* (Princeton, NJ.).

Mangel, M., and Clark, C. W. (1988), *Dynamic Modeling in Behavioral Ecology* (Princeton, NJ.).

Martin, P. S. (1984), 'Prehistoric Overkill: The Global Model', in P. S., and R. G. K. Martin (eds.),*Quaternary Extinctions* (Tucson, Ariz.), 54–403.

May, R. M. (1974), 'Biological Populations with Nonoverlapping Generations: Stable Points, Stable Cycles and Chaos', *Science*, 186: 645–47.

Morse, S. S. (1993) *Emergin Viruses* (Oxford).

Nicholson, A. J., and Bailey, V. A. (1935), 'The Balance of Animal Populations, Part 1', *Proceedings of the Zoological Society*, London, 3: 551–98.

Paine, R. T. (1974), 'Intertidal Community Structure: Experimental Studies on the Relationship between a Dominant Competitor and its Principal Predator', *Oecologia*, 15: 93–120.

Pearl, R. (1927), 'The Growth of Populations', *Quarterly Review of Biology*, 2: 532–48.

Roughgarden, J., Gaines, J. S., and Possingham, H. (1988), 'Recruitment Dynamics in Complex Life Cycles', *Science*, 241: 1460–6.

Roughgarden, J., Pennington, J. T., and Alexander, S. (1994), 'Community Processes at Sea and on Land', *Proceedings of the Royal Society* (London), series B (forthcoming).

Simon, J. L., and Kahn, H. (1984), *The Resourceful Earth: A Response to Global, 2000* (Oxford).

Steele, J. H. (1989), 'The Ocean Landscape', *Landscape Ecology*, 3: 185–92.

Stephen, S. M. (1993) (ed.), *Emerging Viruses* (Oxford).

Terborgh, J. (1989), *Where have All the Birds Gone* (Princeton, NJ).

Walker, B. H., Ludwig, D., Holling, C. S., and Peterman, R. M. (1981), 'Stability of Semi-Arid Savana Grazing Systems', *Journal of Ecology*, 69: 473–98.

Walters, C. J. (1981), 'Optimum Escapements in the Face of Alternative Recruitment Hypotheses', *Canadian Journal of Fisheries and Aquatic Sciences*, 38: 678–89.

Walters, C. J. (1986), *Adaptive Management of Renewable Resources* (New York).

5

'Children are like young bamboo trees': Potentiality and Reproduction in Sub-Saharan Africa

Caroline Bledsoe[*]

Like most great intellectual figures, Thomas Malthus left a puzzling legacy. Depending on which version of the *Essay* we read, Malthus's preoccupation with over-population can brand him as a stern prophet of doom, a testy critic of humanitarian aid, a hopeful child of the Enlightenment, or a naïve, sheltered parson. Nowhere does his credibility evaporate faster than in his dismissal of Sub-Saharan Africa, sight unseen, as a 'savage and neglected state' (1803: 87). Why, then, should a professional cultural relativist be so dazzled by the man?

Malthus laid out a monumental vision of society not as a static entity of fixed parts, but as a vibrant, dynamic system of relationships that reverberate far across time and space (see also Birdsall *et al.* 1979; von Tunzelman 1986). Although he failed to recognize how quickly people can adapt to economic pressures in order to avert grim demographic consequences (see e.g. Sen, Chapter 3, and Holling, Chapter 4, of this

* Associate Professor, Department of Anthropology, Northwestern University, Evanston. This paper, based on research in Sierra Leone, Liberia, and the Gambia, was sponsored by the Institute of African Studies at Fourah Bay College, the Liberian Studies Association, and the Ministry of Health in The Gambia. It was supported by the Ford Foundation, the Rockefeller Foundation, the Mellon Foundation, and the National Science Foundation. Preliminary versions of the paper were presented at the University of Pennsylvania, the University of Wisconsin-Milwaukee, and Indiana University. Particular thanks go to Tommy Bengtsson, Reuben M'Boge, Gudrun Dahl, Partha Dasgupta, C. Magbaily Fyle, Melville George, Susan Greenhalgh, Brian M. Greenwood, Jane Guyer, Eugene Hammel, Allan Hill, K. O. Jaiteh, Momodou Jasseh, Dauda Joof, Ezekiel Kalipeni, William Murphy, Ballah Sillah, Bintou Sosu, Chris Udry, Susan Watkins, members of the reference group assembled by the Royal Swedish Academy of Sciences, and the students in my 1993 graduate seminar on kinship.

volume), he realized that demographic events and their repercussions ripple through time in ways that we cannot perceive by looking only at synchronic slices. A vivid example is his hypothetical case of a mother of twelve who sacrifices several of her sons in war. While the horror of her loss makes society feel boundlessly indebted to her, she eventually causes more misery than happiness because her surviving children place such heavy demands on public resources. Malthus pays keen attention as well to population's chain effects across economic domains: an increase in population acts through the labour market to produce hardships, positive checks, and (eventually) balance, when fewer people result in higher premiums for labour. As Winch (1987: 55) observes, the tension between checks and limits that Malthus detects comprises a 'doctrine of proportions', a search for the optimal conditions, or precise balance, between forces that can produce the best result under changing circumstances.

One of the best examples of Malthus's vision of population dynamics is his preoccupation, in later editions of the *Essay*, with fluctuations in the age at marriage and its potential as a preventive check on population growth. After earlier thunderous prophecies of doom, he became more cheerfully optimistic that moral restraint offered the brightest hope for humanity to take steps through 'preventive' checks to avoid catastrophe. The poor need only be convinced to delay marriage until they could maintain the children they would produce. Underlying this optimism was an implicit assumption that the more humane modes of controlling population could only work when societies became 'civilized'. At this point, 'moral restraint' (what we might translate as refined 'culture') could curb the forces of passion between the sexes (raw 'nature'): 'in modern Europe the positive checks to population growth prevail less, and the preventive checks more than in past times, and in the more uncivilized parts of the world' (quoted in Wrigley 1983: 115).

Malthus's implied evolutionism grates on modern ears. This framework had not yet fully emerged in European social and historical studies. But by the time the fourth edition of the *Essay* appeared, Malthus's synthesis of various travellers' reports of 'wretched', 'savage' societies, and his new interest in delayed marriage as the key indicator of 'advancement', clearly presaged evolutionary styles of thinking. More generally, despite his concern with population processes, Malthus gives us surprisingly little guidance for understanding the family dynamics that generate particular reproductive regimes. As Boulding (1959: pp. xi–xii) points out in his foreword to the *Essay*: 'The sad truth is that although we know a lot more about the mathematics of population than Malthus did, we know very little more about its sociology . . . the subtle social—or even physical—forces which determine the crucial number of children per family.' This failure to explore the dynamics of family organization, coupled with his oddly static—and painfully dated—descriptions of African reproductive patterns, makes Malthus an easy target. But rather than dwelling on his flaws, we can gain more ground by adapting his larger perspective for studying fertility in Africa, where the costs of education, food, and medical care have spiralled, and the complexities of fertility strategies quickly spill over the edges of any household models we try to devise.

In most of Africa large families continue to be valued. Most children, particularly girls, work hard in the household until they leave home for marriage or work, after which many send remittances and ease their parents' material worries in old age. But children have long incurred substantial costs. Even in pre-colonial times, the costs of training, cere-monial transitions, and marriage were enormous (Bledsoe and Cohen 1993). Educating a child for any length of time nowadays strains rural families' scarce income. It also poses an increasing gamble. Because of the shrinking market for wage employment, the chances have diminished that any one child will become highly successful or remit benefits to elders who invested in him or her.

Malthus himself spoke with considerable disapproval of contracep-tion. But if we translated his general premises into today's morals, we might expect people not to dispatch vulnerable children to other families but to mitigate economic hardship by marrying later or by using contra-ception to reduce their fertility. My own research, however, has focused on two widespread patterns that seem to contradict expectations of how families should be responding. These are (1) the widespread use of con-traception for controlling the timing of births, and (2) the widespread tolerance, if not enthusiastic endorsement, of child fostering. Only if we interpret such patterns through the wide-angled dynamic vision that Malthus inspires can we begin to perceive a much larger political spec-trum of worry and action surrounding reproduction.

Malthus's original orienting question, recast compellingly by many of the chapters in this volume, especially those by Sen (Chapter 3) and Birdsall (Chapter 7), continues to pose the key policy issue that the case

I will describe brings to the fore. Should governments enact Draconian measures that may have some chance of stemming population growth but that will almost certainly cause misery in the short term? There is no easy answer to this conundrum. Few serious analysts would dispute the fact that providing food and resources for projected population increments in Africa poses daunting challenges for African economies and environments. As Malthus's work shows repeatedly, however, conventional units of analysis can be deceptive. A narrow regional focus may mask a plethora of global inequalities, externalities, and instances of free-ridership. Western 'near-replacement' fertility strategies, as modest as they sound, still cost the world far more in environmental damage and natural resource-depletion than a narrow fertility focus would reveal, a fact that the north–south debate established some time ago.[1] As such, we confront an awkward conundrum: even if numbers of people were suddenly to stabilize in the world, reducing fertility to two children per woman, raising everyone's level of development and consumption to that of US standards would be ecologically catastrophic.

While these stark imbalances in consumption render fertility-reduction advice to developing countries somewhat hypocritical, the fact that population is increasing rapidly in areas such as Africa cannot be dismissed. Rather than offering a solution to the problem, this chapter attempts to describe in very contemporary terms the nature and extent of the insecurities that make African families reluctant to reduce their fertility. It shows that the diverse skills and social ties that a family cultivates through children are the keys to its ability to cope with the considerable economic adversity and political peril that they face.

This observation has some surprising implications. First, whereas most Western models assume that reducing fertility through contraception is the best solution to economic troubles, rural Sierra Leoneans would find it a drastic solution: most people continue to report in surveys that they want 'all the children God gives them', and typical reactions to suggestions to reduce fertility range from polite interest to outrage. Far more important to them than numbers of children is the spacing of their births. Secondly, while Western models assume that all children in a family incur roughly equal costs, adults in rural Sierra Leone prefer to have a number of children, then invest heavily in the most promising and wait to see what opportunities may arise later for the other children. Finally, by contrast to Western views of rural Africans as clinging to traditional child-raising practices, families in the remotest villages are perpetually attuned to innovative opportunities for children to learn ever-changing bodies of knowledge and skills and to develop contacts with the outside world.

Defending such unorthodox points demands that we venture far beyond ideal functional descriptions, and delve deeply into the micro-political economy of fertility: the dynamics of both mutual assistance and inequality within families. The observation that emerges consistently is that families exert prodigious efforts to control reproduction. Whether such efforts take the form of contraceptives, abortion, breast-feeding practices, schooling, or child-fostering, they imply strongly that nowhere in the reproductive process are people helpless victims either of biological destiny or of traditional norms.

The ethnographic material used here comes primarily from field data on child fosterage collected among the Mende people in Sierra Leone in 1981–2 (unless otherwise indicated, all quotes below come from this research). Other materials come from an earlier field study on marriage in Liberia and a new project on contraception in the Gambia (with demographer Allan Hill). The fact that the material from both Sierra Leone and Liberia was collected during times far less troubled than the present should underscore all the more forcefully the fact that people's concerns with uncertainty as a basis of their reproductive lives are well founded. Though limited particularly to coastal West Africa, the findings point to fresh ways of interpreting strategies surrounding reproduction elsewhere in time and place.

1. MALTHUSIAN FOOTPRINTS IN AFRICA?

With an overall total fertility rate of over 6.1 children per woman and a growth rate of almost 3 per cent per year (Population Reference Bureau, World Data Sheet 1993), Sub-Saharan Africa remains the one world area where fertility has not declined appreciably. By almost everyone's count, children's labour and their potentials for later support still comprise a net gain (see e.g. Frank and McNicoll 1987), and there is little variation in birth intervals, from first to last, that suggest that parents are trying to reduce fertility.

Such intense desires for high fertility are difficult to reconcile with the fact that crushing economic troubles continue to beset the region. These include the demise of reliable formal sector jobs and timely pay cheques for urban workers; the loss of agricultural lands to erosion and urban spread; the inflation of prices for food, medicines, and petrol; and the weakness of national currencies, including the precarious fate of the convertible *Communauté Financière Africaine* in francophone West Africa as the economic map of Europe is redrawn. How, moreover, can we explain Africa's persisting high fertility rates in the face of what Malthus would interpret as drastic 'positive' checks (famine, wars, disease) on population pressure, and only mild rises in the more benign 'preventive' check of delayed marriage? Such issues are by no means simply academic ones. Africa stands at a crossroads in the eyes of the development and population world. Should governments and donors take forceful measures to limit growth rates and plough disproportionate investments into contraceptives? Or should they settle for *laissez-faire* strategies of health and development, hoping that the hidden hand of reproductive reason will eventually prevail?

Although Malthus was clearly obsessed with problems of over-population, a major surprise emerges from reading his book through Africanist eyes: the close convergence between his ideals of reproduction and those of most contemporary Africans. For both sides, the primary purpose of marriage is propagation. Once marriage begins, reproduction should also commence, with no attempt to regulate numbers of children through contraception. Malthus, in fact, spoke with more indignation about controlling child-bearing within marriage than about bearing children before it. Finally, his position that people should not marry till able to support a family did not preclude high fertility; he saw some population growth as a healthy stimulus to the economy.

Yet the parallels can be pushed only so far. While Malthus disapproved of reproduction before marriage, many African societies tolerate varying degrees of sexual expression among unmarried people, though they prefer that children have a father who recognizes and supports them. (A responsible father is considered more essential to a child's welfare than the mother's marital status.) Further, while most African families welcome high fertility, Malthus's 'water-tap' image of marriage unleashing a torrent of fertility is a dim reflection of African reproductive realities. Birth intervals are a source of considerable worry: short, uncontrolled intervals are feared to endanger the lives of mother and child. Instead, births are ideally separated by generous intervals of at least two years, and many rural women maintain exceedingly regular birth intervals of two years or more. Further, whereas Malthus, despite his denunciation of contraception as an 'improper art' that aids and abets vice (1803: 18), saw such devices as potential checks on fertility, many African women have enthusiastically adopted contraceptives not as devices to reduce children, but for spacing births, a point to which we will return later.

The widest gulf separating Malthus from African reproductive realities, however, is the one that will concern us here: the management of children's costs and benefits. While Malthus argued that delaying marriage was the best way of ensuring responsible fertility in situations of economic hardship, most rural Africans remain intensely anxious to maintain high fertility, regardless of economic circumstances.

The Costs-and-Benefits Framework

We can take the first step towards explaining African deviations from Malthusian expectations by looking more closely at three questions that would fall under a standard cost–benefit rubric. These questions ask how the costs of children are absorbed through time, space, and social networks.

1. *Who pays for, and benefits from, children?* In Africa, children are cast as the common property of extended families, and their costs and benefits are distributed through social networks that range far beyond the biological progenitors (see e.g. Isiugo-Abanihe 1985; Caldwell and Caldwell 1987; van de Walle and Ebigbola 1987; Lesthaeghe 1989). A child's advancement is viewed as a means towards the larger end of helping a much wider social group; in the succinct words of a Mende proverb: 'A child is not "for" one person.'

Few parents can pay all their children's costs, many of which inevitably arise during crisis. To fill in the gaps, anyone from rural grandmothers to patrons, schoolteachers, or urban kin can be asked to help with expenses. They can also foster children to train them in new skills and help them develop outside contacts (see e.g. Oppong 1973; Goody 1982). Among the most important supporters of children are their successful elder siblings. In turn, these children are expected to support their yet-younger siblings, and the children of the older sibling who supported them, in a chain of obligation. 'Child spacing', therefore, is viewed locally not simply as a means of promoting children's health; it is also a means of more easily shifting the responsibility of raising and educating younger children on to the older ones. A man who regretted that his own children were closely spaced described the ideal situation:

Say if my first child has started school since 1979, then I am expecting to send the other one in 1985. Then I would expect that the five or six years would help me because by the time the other one finishes the secondary school, the other one might be starting, and I expect the moment you [the older child] complete your secondary school, you would engage in some other work to earn you money. So that would . . . enable the first born to help in educating the others.

The ideology that declares older children responsible for their parents' fertility lightens considerably the parents' own economic consequences of high fertility, at least in principle. Conversely, while economic problems may induce people to seek help in raising their children, the knowledge that help for raising children is available dissociates childbearing from immediate economic worries. A young man's testimony clearly articulates this relationship:

If you have a sister, a brother, or an uncle, you can send any of your children to them, and they cannot refuse them. So you know ahead of time, even when you have one or two children at first, that if you have five or six more, you can send them out. We are not worried about taking care of [children]. We just know that every child can be raised in a home, whether [with] the born family or the relatives. That's why we have the extended family.

But, while parents do not bear all the economic responsibility for their own offspring, everyone contributes to the costs of everyone else's children, whether willingly or not. Even adults with children of their own can be asked to help pay the expenses of other people's children.[2] In addition, children forage outside the boundaries of the household, running errands for tips, stealing from gardens, and embezzling market funds and school fees (Bledsoe 1991).

2. *Do all children cost the same?* Children in Africa, even siblings within the same family, often incur very different costs as families try to diversify their skills and networks.[3] Since families may spend substantial amounts on a few children but bring up most at subsistence level, the result is a long continuum of support, ranging from enormous investment to marginalization. (See Eloundou 1993, for an analogous argument. For

important discussions of intra-group inequalities, see Sen 1981 and Folbre 1986.)

'Cleverness' marks an important line of potential differentiation. Some 'clever' children may attract substantial investments for schooling or advanced training in Islamic scholarship; others are put to work as soon as possible. A bright student who excels in the General Certificate of Education exams will have no trouble mustering support from individuals (related and unrelated) who are eager to help with further educational expenses; a primary school leaver, by contrast, finds it difficult to enforce the most minimal lineage rights. Similar patterns apply to gender; because a poor family is reluctant to expend limited resources on a girl who may get pregnant and be evicted from school, girls face imposing obstacles to advancement. An older woman made clear the implications of these patterns from a mother's perspective:

Girls start being an advantage to you and helpful to you when they are matured and join the Bundu [initiation] society and are given to a husband. Then she [a girl] can benefit you . . . When my first daughter was twelve, I gave her to a husband and started getting benefit from her. She wasn't educated . . . I was praying to God to first give me a girl because I can see her benefit soon . . . With boys, it depends on the education. At about the age of 21 and over, when the boys might have completed schooling, picked up jobs and started earning money, they will be able to give support to the parents.

Another domain of distinction among children stems from their mothers' statuses. Despite ideals of egalitarianism among the co-wives of a polygynist, the children of a low-status wife may be dispatched to trade apprenticeships or 'home-training', while the household budget is strained to educate the children of a wife from a high-status family. And paternity clearly can differentiate children. By denying that he fathered a particular child, a man can try to avoid further financial obligation. Yet most cases of denied paternity represent less a permanent lopping-off of a child than a man's deliberate effort to evade paternity responsibilities until subsequent events unfold. When a child wins a scholarship to college, a man who previously expressed doubts that he fathered it may suddenly discover new evidence that he is in fact the father.[4] My own project actually produced an analogous case.

I first met Seiku, the young man who became my main assistant, in a small regional outpost of the Central Statistics Office, CSO, in eastern

Sierra Leone. He was a full-time civil servant in the CSO, employed as a surveyor. Since there was no major census or development survey in sight, the CSO was happy, even eager, to second him to me. Although reluctant at first to leave the security of his office and venture into a small rural town, Seiku quickly realized that he was about to become much more wealthy than he would through the humble pay-scale of the census office. He began to order tailor-made clothes, buy imported leather shoes, and send his mother small gifts of money and food. Seiku's mother and father, I learnt, had been divorced since he was a child. He had heard little from his father since then; repeated requests from his mother for money to help support the children had fallen on deaf ears. But when his father heard (as one inevitably would, through the extensive rural grapevine) that Seiku was working for an American, whom everyone knows must be laden with money, his father sent him a note suddenly resurrecting his status as father and demanding a huge amount of money to plant a coffee plantation.

3. *When must children's costs be paid?* One of the ways in which adults try to maximize children's potentials by wise investments is by spreading costs through time. As a result, they can try to continue bearing 'all the children God gives them' by investing heavily in the most promising children and delaying substantial investments in the rest.

2. METAPHORS OF POTENTIALITY: CHILDREN, BAMBOO TREES, AND 'DEVELOPMENT'

Taken alone, these facts should undermine our confidence in traditional formulas for assessing the costs and benefits of children. But if we left the argument here, we would still have only a superficial grasp of the larger political culture of reproduction. A young man's statement makes the point with simple elegance: 'We are interested in having children. It is very good to see them. They are children, and I will not try to stop giving birth to them.' While attempts to link fertility patterns within Africa to various economic types may yield mildly positive results and rates of child-bearing may vary slightly, the more important fact, that births continue at evenly paced rates and virtually independent of economic regimes, gives us strong grounds to suspect that we may be missing the forest for the trees. Rural people do not seem to be attaching costs to children as Western families might; insecurity and uncertainty make them immensely wary of gambling with something so fundamental as reproduction.

Asking how families try to cope with pervasive economic uncertainty and political insecurity, Wolf Bleek's (1987: 148–9) vivid explanation of why married women in the Kwahu Plateau of southern Ghana do not actively plan their numbers of children provides clues to a wider vision of reproduction:

The context in which women give birth to children and bring them up is subject to so many vicissitudes and their conjugal situation is so uncertain that their ability to foresee their future is extremely limited. For a woman, it is extremely hard to estimate the pros and cons of having a few or many children. It depends on how long her marriage will last; how many times she will marry; the financial position of her husband(s); what conjugal responsibilities her husband(s) or lover(s) will accept; how many of her children will be staying with her; how many children of others will be put in her care; how much help her lineage will give; how successful she will be in earning her own income; how much help her children will offer; how successful her children will be at school and in achieving a good economic position; how healthy her children will be, and how many of them will survive to adulthood. Of course, some of these uncertainties are universal, but it should be taken into account that in the relatively rich industrialized societies at least the worst of the economic pressures resulting from misfortune have been greatly reduced by social security and insurance measures. In Akan society such measures hardly exist. The greatest unknown there, as in many other Third World societies, is that children can become factors of social insurance as well as factors of social risk. The aggregate 'decision' in such a complex and contradictory situation is most likely inertia. Inertia is also brought about by the strong traditional norm of high fertility. Children are the *raison d'être* of marriage. Kwahu people engage on the uncertain course of marriage because they want children. So to stop or to limit fertility while married seems contradictory and has a tinge of absurdity.

As we attempt to move beyond the cost–benefit rubric and into this more experimental turf, a provocative Mende proverb sheds light on a very different philosophy about children, one that emphasizes potentialities. 'Children are like young bamboo trees; you don't know which of the shoots will be cut away, and which will remain.'

In the wild, a bamboo tree begins life as a clump of small green shoots that rapidly thicken and thrust their way upwards. The maturing shoots send out yet more colonies of shoots, and all compete fiercely for space and light. Wild bamboo trees, in their maturity, are enormous masses of tightly wedged green trunks surrounded at the base by piles of dried poles that were crushed out of the way by their stronger brethren. Bamboo trees that have come under human control are quite different. A bamboo tree that happens to grow near human habitation is subject to constant pruning as people trim away its maturing shoots to use in fences. Only a few are allowed to grow thick and tall for more substantial uses later.

If read too literally, this particular proverb might imply that children's lives, like those of bamboo shoots, may be terminated intentionally when fertility is too high. But when read more figuratively, the image implies that children are creatures of unknown potentials. Some may die, one or two may achieve great success, but most will achieve very modest goals. Since people are unsure which children will do what, they try to avoid making pivotal decisions about either fertility or investments in children that would forever commit them to one course of action or another; rather, they try to retain options as long as possible. Instead of trying to avoid having children, a wiser course is to have as many children as God gives you—but to take active efforts to assess children and shape their prospects *after* their birth. Substantial investments can be delayed until more information is available, as children's potentials unfold.

The cost–benefit rubric, then, provides a useful starting-point because it gives credit to even the most underdeveloped, isolated people as rational decision-makers, given the amount of information they have. But cost–benefit analyses assume complete, or almost complete, knowledge. The notion of potentiality, by contrast, suggests that not only do people know that their information is incomplete; it also suggests that they know that new situations will emerge which in turn will present new opportunities and new constraints. Such a perspective is vitally important to decisions about children, whose identities, talents, and powers emerge only over time. (See also Willis, Chapter 6 of this volume.)

One source of potentiality in children is their own inherent talents and capabilities. Ideally, a family produces a range of strong, knowledgeable, skilled members. A rural man elaborated the virtues of such diversity:

It is important to have many children. First, to help the younger ones, and secondly, to help you at your old age, so when you are old, you will have support coming from different angles or sources. Let's say your first son sent you a bag of rice, you have a daughter who is a nurse [to] provide medicine, another child is a Permanent Secretary who will send money for you to maintain the house, and so on.

But children's potentials are not always 'good'. A Mende proverb articulates this realization: 'A snake does not have a child.' That is, one should not treat children lightly or underrate their potentials. While young, they can be dangerous; when grown, they can wield powers that may, in the worst case, destroy the family. A child with a bad character may infuriate political authorities or embroil the family in expensive lawsuits. He can even be an evil spirit that has come disguised as a baby, with the specific intent of destroying the family from within. The need to produce children of good potential places special pressure on women, whose behaviour (and even the food they eat) is said to determine a baby's character. It is little wonder that the Mende regard a pregnant woman as being in *kiti-hu*, a dangerous state of moral and physical ambiguity, and her new-born with considerable ambivalence until its character begins to emerge.

Other people's children can also have certain potentials for one's own family. A bright child in the village may bring handsome rewards to those who invested in him when he goes on to become a national leader. Conversely, the success of a child from one family may spell the downfall of its enemies; his victory in a chieftaincy election can destroy rivals' access to land, jobs, scholarships, and legal assistance.

Children, then, have their own inherent potentials, whether good or bad and in the eyes of one family or another. Equally important, the outside world is a place of potentials where children can gain knowledge and power: a process the Mende call 'development'. By child 'development' I do not refer to the Western notion of physical or emotional growth. To the Mende, 'development' means 'progress' (*tee-guloma*: literally, 'to pass in front of' or 'to pull out'). It refers to the act of shaping something into an entity more advanced than if it were left to take its natural course. 'To develop' can also be used as a transitive verb, referring to the act of 'developing' *something*—a household, a town, or a country, or even a child—so that the benefits derived from it are greatly multiplied. Nowadays, a 'developed' child has been to school; lives in a modern house; wears clean tailored clothes; speaks English; entertains important visitors with 'civilized' foods; works in an office; and earns a monthly salary, instead of relying on irregular seasonal harvests.

The fact that families in Sierra Leone need to stay connected to diverse, ever-changing sources of power and economic stability begins to erode our confidence in crisp cost–benefit formulas. Adults cannot possibly know what new possibilities will emerge, once a child is born. A child, himself having unknown potentials, will find himself in places that offer unanticipated opportunities, and he will come in contact with people who themselves have obscure potentials. The value with which a child begins life will therefore change dramatically, depending on the opportunities he encounters, the contacts he is able to make with other people, and even the opportunities and contacts that these other people themselves develop.

3. THE POLITICAL ECONOMY OF REPRODUCTION

Westerners tend to separate issues of children from politics, an effort stemming, perhaps, from a distaste for assessing children in pragmatic ways. We usually relegate the topic of children to socialization or what we perceive as domains of uncomplicated amity: the 'domestic' sphere or 'kinship'. Material from Sierra Leone, however, leaves little question that the domestic domain is anything but uncomplicated amity, and reproduction and socialization are inextricably linked to patronage politics.

In Sierra Leone, as in much of Africa, land cannot be owned or sold privately, but individual family groups lay usufruct claims to it. Rice, the staple, is cultivated in upland plots by slash-and-burn techniques involving heavy labour inputs at peak seasons and the rotation of land plots (see Richards 1985). Since the turn of the century, rising population pressures and incentives to engage in cash-cropping have led to reduced fallow periods, soil erosion, labour intensification, and the alienation of land from subsistence cultivation. People have migrated to wage-labour jobs or to urban areas, but the uncertainty of these jobs and regular disruptions in pay cheques make everyone maintain a keen competitive interest in the rural areas. For rural people, hard times can be disastrous. Growing scarcities of food, money, and petrol and the collapse of public services give people no choice but to cultivate ties to myriad networks of people who may, at crucial times, have access to scarce resources.

Equally worrisome are political dangers and uncertainties. While the modern system offers enticing opportunities for wealth and advancement, jobs, scholarships, and legal muscle trickle down through personal ties to powerful brokers who can intervene with national institutions. Court cases that can strip people of land, property, and dependents are commonly trumped up against those with weak patronage support. Illiterate rural people are particularly vulnerable to heavy-handed government officials who invent taxes to collect and demand 'contributions' to fictional building projects. In effect, low-status people must acquire their own 'big men' to protect them from other 'big men'. Precipitous declines in the national economy make it imperative to have both well-connected patrons and diverse networks, to bypass bureaucratic road-blocks during inevitable shortages of food, money, and petrol, and to provide access to the international world of mobility, jobs, and hard currency. Wealth and security, then, lie fundamentally in people and relationships, especially those with the outside world, far more than in money or property. Under such precarious conditions, bearing healthy, clever children and training them for success are fundamentally political acts.

The Political Economy of Child 'Development'

For women, bearing and raising children is the heart of security. Guyer (forthcoming) points out that anthropologists and demographers alike have viewed fertility in Africa in lineage terms, as a source of future security and labour some decades after their birth; yet the need to build lateral ties to other groups in an increasingly unstable political economy is leading women to place more emphasis on children's potentials for creating immediate links to men and their networks. In the Gambia, for example, numbers of children are less important to a woman than whom she has them by, and whether she is having them regularly as a manifestation of conjugal loyalty. Similarly, Guyer shows that a Nigerian woman has far greater leverage to press her demands on a man, whether or not she is married, if she has a child by him. Since unmarried women usually keep children with them even if they do not hold formal lineage rights in them, some women find it advantageous to bear children by different men to gain access to multiple resource networks. Although a woman cannot marry polygamously, she can have several concurrent fathers to her children, resulting in what Guyer calls 'polyandrous motherhood'.

Children comprise the heart of security for families more generally. For some children, families are content merely to 'raise them up' to take over the home and farm, a task that can be consigned even to an illiterate old 'granny'. But other children have the potential to strike out into the world beyond the home and farm, to become something *other* than what the family now is. Since people need ties to patrons in the outside world who can ward off exploitation and provide hard currency in times of need, they constantly watch children, trying to discover and cultivate their potentials, in order best to deploy them to the outside world. A rural 40-year-old woman interviewed in our current Gambian project was quite explicit about the importance of ties to the outside world through children: 'my co-wife . . . has five children, all of whom are living in the outside world . . . She receives very good support, financially, from all her kids regularly. This is why she is better off than me' (field notes collected by Fatou Banja-Camara, September 1993). The central point about children and change needs underscoring. The classic African kinship literature has stressed that, although lineages sought to forge political ties to outside groups through marriage and political alliances, they were essentially inward-looking. They drew on internal resources and had reproduction of the old as their basic agenda. Family elders were seen as conservative forces seeking to uphold traditions and reinforce ties to ancestors. But, in actuality, the family that remains the same is doomed to extinction. Children, more so than any other kinds of people, are expected to stake entrepreneurial claims in rapid flows of change, precisely because they are poised for training at the time when new skills are becoming available. Though families may hearken nostalgically back to the old days, they must be ready to capture the benefits of these wider changes by producing a constant supply of children who can continually break new ground within the new opportunity structures that emerge— contexts in which family members themselves may have little expertise.

Paying ritual initiation fees has long comprised a symbolic attempt to lay future claims on children with great potentials. More informal means consist of inviting children to eat when they are hungry, or even contributing to the funeral expenses of a relative. An increasingly important strategy is paying the school fees for children known to be clever, for one of them may become a chief or even a district officer. Paradoxically, then, rural villagers, who are usually considered to be most ill at ease with intimidating modern institutions such as education, are most anxious to invest in them. A man explained:

Now in this chiefdom we are lacking education. So we want to assist the small children; whether it is our own child or our friend's child, we assist them. Good will come out of this, whether it is our own child, or it is not our own child, just as long as he [the child] is clever . . . If I spend my money on those children who are bright . . . those children . . . will go further in education and later they may do something for me that will make me to forget about all what I am spending on them now. The reward I get may be more.

Child 'development' comprises a key strategy of social mobility in a hierarchical system based on ties of dependency. In the ideal of family development, each generation can theoretically boost the family up to the vantage-point of its most successful members. Using each new level as its 'anchor' (as one young man put it, using an English metaphor of mountain climbing), the family can then 'develop' further by sending out new children to traffic with ever-more advanced levels of society.

One very rare kind of child looms large in Mende minds: a child believed to be specially blessed by God and the ancestors who has the potential to become a leader and help the extended family as a whole to 'develop'. Given the right training, this child will quickly supersede the humble surroundings into which he is born. He will study hard in his tiny village school, earning the praise of teachers who will facilitate his admission to a more prestigious regional school. He will defeat scores of competitors to win an overseas scholarship and, eventually, return to win a seat in the national parliament. From this vantage-point, he will use leverage with the minister of education to obtain scholarships for his relatives' children, and he will persuade his new friends, the officers in international development organizations, to establish projects in his family's town that will bring running water, electricity, schools, and employment. Such children may be born to a family only once in a lifetime, if ever. Most families consider themselves extraordinarily blessed if a son acquires a job with ready access to hard currency, or a married daughter living in Europe sends for her mother during difficult times back home.

Needless to say, simply paying children's costs will not automatically produce children like this. Nor will 'developed' children necessarily bring future benefits at all. Among the most tragic kinds of cases of child 'development' are those children who fail to measure up to their potentials, drop out of school, and get evicted from the guardian's home for defiance. Such children will not only fail to benefit their families; they will sap resources. They will hover endlessly around their better-off kin, pleading for help with the most basic necessities. Even more tragic are the talented children who consume all the family's resources in the course of becoming 'developed', but then delude themselves that they earned their successes independently through their own hard work, and turn a blind eye to the needs of their benefactors. (Successful children, of course, see the problem quite differently: they find it utterly impossible to satisfy every demand from dependent kin, all of whom now claim to have been instrumental in their success. Trying to cope with the expenses and uncertainties of urban life, they may resort to short, unannounced visits to their rural homes or feign sickness during visits to discourage long lines of 'benefactors'.)

Older anthropological works led us to believe that certain relationships entailed clear obligations. But we now know that people, like objects of value, have contested histories of ownership, transactions, and exchanges (for related discussions, see Comaroff and Roberts 1981; Appadurai 1986; Kopytoff 1986). In Sierra Leone, claims to, and obligations towards, children comprise a tangle of fragments that are renegotiated constantly over the years. Adults must continually shore up the claims they allege children owe them, reiterating their rights and exaggerating their past contributions to the children's upbringing.

Fosterage is one of the most striking examples of how people hope to gain security from 'developing' children. Fostering in the African context is emphatically *not* adoption; it is not intended to break ties between parents and children. To the contrary, fostering is usually regarded as a highly positive opportunity for older children to obtain knowledge, degrees, and contacts that will break ground for the family into a number of diverse new occupations and networks.[5] Since parents lack the resources to 'develop' their children much beyond their own socio-economic level, they try to send promising children out to live with guardians in the 'civilized' world. These children often undergo considerable hardship in their foster households: harsh discipline, food of poor quality, and so on. But while young fostered children appear to be at greater risk from morbidity and mortality than those with their own mothers, they may actually be at less risk than if they had stayed with their mother, especially if she has no stable conjugal partner. And while fostered children being raised by a high-status family are often denied its richest resources, they may enjoy better facilities than if they had remained at home (Bledsoe and Brandon 1992).

Family 'development' is the ultimate goal of parents' fostering motives. In the ideal, high-status guardians, seeing talented children's potentials, send them up to their own patrons, and so on, in a step-migration model of ever-greater academic and geographic prestige. Ministers of parliament from rural areas who overcame enormous hardship become part of national folklore and models for aspiring children to emulate. In contrast, then, to the firmly held Western belief that children need stable parental figures throughout their childhood, most Mende would argue the opposite: the truly unfortunate children are those who have *not* been sent away from home to advance themselves.

Because fosterage sets up and symbolizes patronage ties, fosterage strategies of mobility for children follow lines of social and geographic stratification. Throughout the polity, parents try to send their own older children to more-urban areas and to people of higher social and educational status, while simultaneously taking in children from more-rural areas or households of lesser status, whose parents seek similar opportunities for them. (My 1982 survey in Sierra Leone found that 51 per cent of all fostered children under the age of 16 were in higher-status homes than those of their natal families, compared to only 17 per cent in lower-status homes. Thirty-two per cent were in same-status households.) At each upward step, the children's lower status marks them as domestic servants for more educated or urban families. The greater the status differential between the sending and receiving families, the less the foster children will be treated like the guardian's own children, and the more hardship they are likely to undergo. The trade-off, of course, is the possibility of an incremental rise in the child's adult status. Such patterns bear a striking resemblance to descriptions of highly stratified societies in historical Europe. In Tudor England, 'Each rank, even noble ones, could hope that the experience gained in a home superior to their own would give the child a sophistication, breadth of outlook, and opportunity for advancement that they themselves could not provide' (McCracken 1983: 311).

Besides the benefits that children themselves gain from fostering, they and their families also gain political ties to important guardians. In Liberia, Americo-Liberians living in Monrovia were regarded as hostile, punitive autocrats, at least until their bloody fall from power in the early 1980s. But because they comprised the dominant political force in the country, indigenous families often sent promising children to be their servants in order to gain entry into the top crust of national patronage circles. Although the ethnic identity of the élite has changed—or perhaps *because* it has changed so swiftly and brutally, producing vengeance and terror at all levels down the hierarchy—mechanisms such as fostering should not be interpreted simply as distressful responses to the economic hardships of high fertility; they are a political imperative.

Fostering holds clear benefits for guardians as well. Because of the extraordinary 'development' opportunities he is said to have gained by virtue of the generous guardian who took him in, a foster child is under heavy moral obligation as an adult, as a local saying stresses: 'There is a very great debt on you to the one who raised you.' Since a successful child and his descendants will forever reap the benefits of his enhanced 'development', those who invested in him consider their contributions not as loans, repayable as finite sums, but as investments whose proportionate values must be repaid perpetually as the beneficiary's earnings multiply. A prominent Freetown lawyer I knew reported that, although his former benefactor—the man who fostered him, paid his school fees, and sent him to England for advanced studies—is now dead, he continues to show his gratitude by putting the man's children through school. With an eye to such success stories, most people foster a number of children throughout their adult lives; it is rare to find adults who have never fostered children. A man explained why, in a manner that recalls the bamboo proverb: 'It is good to bring up or mind [foster] a lot of children because you don't know which one will be successful. However it may be, one *must* be successful, and you will get your reward out of your expenditure.'

4. RETAINING REPRODUCTIVE OPTIONS

The constant struggle to cope with shortages, uncertainties, and exploitation by maintaining multiple, ever-changing sources of power and stability helps to explain why rural families in Africa try to keep their fertility options open as long as possible to make the best investment decisions in the face of uncertainty. This knowledge allows us to discern the contours of distinctly African responses to a wide corpus of social arrangements and technologies that offer possible solutions to potential high-fertility problems. The categories that begin to emerge are those involving giving up children altogether versus holding on to as many children as possible by reducing the costs of some.

Holding on to Children

If US parents were asked how they control the costs of fertility, their answers would be likely to involve hormones, barriers, surgery, and uterine implants—a set of medical solutions that can be used to limit births to a 'wanted' number. Yet the actual possibilities for controlling the costs of children are far more extensive, once we open our eyes to all the activity that can occur *after* a child's birth. Far more culturally preferred in Africa than strategies that irrevocably reduce numbers of children are those which allow parents to hold on to children but, if necessary, allocate resources differentially among them—delaying resource allocations for some children while promoting the 'development' of the most promising.

Fosterage is the most obvious example of these 'holding-on' strategies. Descriptions of fosterage in Africa depart sharply from what Hajnal (1982) discovered was the case for pre-industrial North-West Europe: young people's need to serve in other households in order to amass the resources to marry produced *low* fertility because marriage was effectively delayed. Unlike Hajnal's servants, most of whom appeared to be of marriageable age, foster children in Africa are usually pre-pubescent, and people inevitably consider fostering as a form of support for *high* fertility. Fosterage, unlike contraception or abortion, leaves enormous scope for subsequent manœuvre. Fostering arrangements can be made and unmade a number of times, even with the same child, to meet new exigencies. At a broader scale as well, fostering dramatically reflects cost-balancing mechanisms: urbanites can shift the costs of feeding and caring for small children to non-cash channels in rural subsistence areas until the children are ready for better schooling or specialized apprenticeships that are best obtained in urban areas.

Disproportionate investment—in its extreme form, marginalization— is another alternative to reducing numbers of children. The conviction that the best way to ensure the 'development' of families in a climate of instability is to invest differentially in children frequently means that some children advance quickly while others become marginalized.

Defending their actions, people stress that marginalization is potentially reversible, since the child can be reincorporated if circumstances change. They also insist that such children, like other members of the family, are potential beneficiaries of highly successful children. Reality, of course, may be quite different. But, from the point of view of the family, and certainly for marginalized children themselves, allocating resources unequally among children is far better than preventing their births.

Giving up Children

If the possibility of relinquishing some children altogether is most antithetical to African cultural logic because it cuts off potentials most quickly, then it should not be surprising to learn that most of the options that involve giving up children altogether are extremely rare in Africa. Equally important, the apparent exceptions seem to prove the rule.

Celibacy is an extreme case, one almost never heard of in Africa, except among those rare individuals who join Catholic orders; in fact, bans on reproduction probably constitute the major hindrance to Catholic recruitment. Another extreme case is sterilization, done only rarely in Africa. Yet our work in the Gambia has shown that, while a number of older, high-parity women desire sterilization and a tiny few even manage to obtain it, reducing fertility is not their chief motive. Since maternal mortality is quite high among such women (longer, more difficult labour; increased risks of haemorrhage; and so on), avoiding medical risk is by far the stronger motive.

As for infanticide, Scrimshaw (1978), writing about Latin America, describes this as a way of reducing family size for poor couples lacking means of birth control. To be sure, infanticide strategies contain some subtleties that a simple 'fertility-reduction' argument obscures. Skinner (1993: 252) reports that, in Tokugawa Japan, infanticide was preferred to abortion because it was safer for the mother and because it created more space between those children who were allowed to live. He also points out that infanticide seems to have been practised both to pare down numbers of less desired daughters, and to delay the arrival of the much-prized male heir until a grandmother or an older sister was available to pamper him. In contemporary China many cases of infanticide result from the forceful one-child-per-couple state policy that mandates late-term 'abortions'. But it also serves 'pruning' functions in the one-child context, when parents must wait till the baby is born to decide whether they want it. Among parents anxious that their 'one child' should be a boy, eliminating a first-born girl effectively wipes the slate clean (Greenhalgh, 1994). (In India, the much-publicized increase in aborted female foetuses after the advent of amniocentesis technology is a variant of this case; see e.g. Miller 1987.)

But, however common infanticide might be as a solution to fertility imbalances in Latin America and Asia, it is quite rare in Africa, outside the few remaining hunting-and-gathering groups and apart from cases of perceived physical or spiritual abnormalities. The reasons probably stem from the fact that infanticide, by ending a life, represents an irreversible measure. Among cases of infanticide, the reason why a child was maltreated or neglected seems related more to deteriorating relations between the parents than with economic desperation. In a recent case reported in the Gambian *Daily Observer* (1 Oct. 1993), an unmarried 22-year-old woman dumped her two-day-old baby into a pit latrine because the father 'apparently ignored her and [refused to] share responsibility for the baby, which she felt was a burden on her'.

Cases of child neglect may be more common than the more dramatic instances of infanticide. Since women need children as links to male resources, women with new partners may feel pressure to shore up these new unions by displaying a lack of interest in their own children from old ones. An insightful man explained:

When the woman gets the child with her new husband, she would want to satisfy the husband by taking good care of his child. He would in fact be vexed to see her taking care of the previous husband's children so well, and there will then be misunderstanding between the new couple, and the woman wouldn't want that. So it is not that the woman loves her older children less, but that she knows she must take better care of the new ones to make sure that no palaver [discord] arises between her and the new husband.

Adoption, in the sense of a permanent legal transfer, is reported commonly in world areas where people have too many mouths to feed. In Oceania, where island geography places clear limits on land supply, ethnographers have long argued that adoption is used widely among extended kin to gain access to scarce land and to balance fragile populations in the face of severe weather, epidemics, and the sways of demographic chance (see e.g. Fischer 1963; Carroll 1970; Brooks 1976; see, however, a caution against this interpretation by Keesing 1970). In Africa, though, the English word 'fostering' is usually closer to what people mean when they give a child to someone; parents usually monitor the child's progress and welfare, and few people would argue if they eventually took back the child. In those few cases I encountered where people truly did not seem to want the child back, the child seems to have been given symbolically to tighten an already close bond between families. Although giving up children for adoption in order to reduce economic burdens very likely occurs elsewhere (cf. Knodel, Chamratrihiron, and Debavalya 1987, for Thailand), it would be wrong to interpret most cases of adoption or even fostering in Africa as manifestations of an unmet need for contraception. Fostering out a child, even to conditions of abject rural poverty, is much preferred to giving up a child altogether.

Across the continent, abortion is difficult to document because of its illegality, but reports of it arise most often among married women who fear condemnation for short birth intervals or illicit pregnancies, but especially among schoolgirls who fear they will be evicted from school (Caldwell and Caldwell, forthcoming). Although some cases of abortion may stem from pregnancies that were unwanted from the outset, this is not always true. Abortion has an important advantage over contraception. In Africa, pregnancy, much like marriage, is often less a discrete event in time than a lengthy process. Having abortion as a possibility gives a woman time within the pregnancy process to reflect on her options. Rather than lose a good husband prospect, a schoolgirl, for example, may decide to proceed with the pregnancy and bear the child if the man's reaction is positive; if he proves uninterested, she may try to abort.

Child abandonment inevitably conjures spectres of foundling homes, baby farms, and street waifs in historical Europe (see Boswell 1987). The practice is increasing somewhat in Africa, especially in the wake of AIDS. But even among those children who are abandoned, many do not seem to be lopped off as severely as reports of child abandonment in Europe indicate. They continue to keep track of their families, and vice versa. Nor are children always abandoned because of poverty. As Guyer (1991) points out, Marguerat's (1990) material on street children in Lomé, Togo, suggests that many of these children, who are actually from middle-class families, were not abandoned but left because of quarrels with step-parents.

The use of Western contraception is the most striking exception that proves the rule that fertility reduction is the strategy of last resort. The use of contraceptives could well be interpreted as a drastic means of controlling the costs of children because it very decisively prevents a life before it even begins. Yet in the Gambia, our study is showing that many rural women are eagerly adopting Western contraceptives; but not for the reason that most of the industrialized world assumes: to reduce fertility. Instead, they are incorporating Western contraceptives, especially pills, with more long-standing methods of spacing births to place finer control over the timing and circumstances of births.[6]

One of the most important pieces of evidence that rural women use contraceptives for purposes other than fertility limitation is again found in the exceptions: those who use contraceptives but do not appear to have spacing in mind. These are almost exclusively older women, over the age of 35, and they are using almost exclusively the injectable Depo-Provera. In theory Depo lasts three months, but it often has an unpredictably long spill-over period that young women, who have only begun child-bearing, find alarming. Like young women, older women do not appear to be trying to reduce their fertility. Rather, they fear the spiralling health risks of high parity child-bearing: ecclampsia, prolonged labour, and haemorrhage. As such, the 'unintended consequence' of the technology, a long, hormonally enforced infecundity, is precisely what they are seeking.

Further support for the notion that people are trying to retain fertility options may be discerned in a different kind of exception: actual fertility reduction in three African countries, Kenya, Botswana, and Zimbabwe (Brass and Jolly 1993). It is, no doubt, significant that these countries have among the most wealthy, but also the most stable, economies in the subcontinent. What is most striking, however, is the anomalous pattern by which the decline has occurred. By comparison with fertility declines in most world areas, which stemmed from reducing fertility in the later parities as women reached their fertility goals, declines in these African countries have come about chiefly through wider spaces between children at all parities. No one is sure why this is the case. Wider spaces, rather than a small cluster of closely spaced children, may reflect families' efforts to reduce the numbers of children enrolled simultaneously in expensive educational institutions. But the spacing, rather than stopping, also likely reflects families' efforts to maintain wide age gaps that can facilitate support for younger children from older siblings, and to stay poised to tap ever-emerging new sources of power and knowledge.

With these few exceptions, forsaking children completely is still an exceptionally rare way to manage the costs of children in Africa. Reducing fertility means that the unique personality and talents of an unborn child are lost forever. That child might have been the one to step forward to assume leadership and avert a family crisis. Families, even in the most dire circumstances, seem to close off reproductive options only as a last resort. Knowing they will have considerable scope for shaping their family size and composition as subsequent events unfold, most people are eager to bear children. It is less the case, then, that people are poor because they have many children; rather, *because* people's economic and political situations are so unstable, they need numerous children.[7]

5. DIVERSITY AND ADAPTABILITY: THE HEART OF THE MATTER

How to balance people's numbers against available resources has been the major point of debate throughout the history of population studies. Besides Malthus, another prominent analyst who faced the issue was Chayanov (1986). Writing in the early twentieth century, he observed that a typical peasant household undergoes enormous changes in the ratio of consumers to workers over its life cycle. Peasants try, nevertheless, to maintain a ratio that is as flat as possible by reducing births, working longer, or acquiring more labour when needed. However, the range of possibilities that both Malthus and Chayanov presented for coping with resource limits was quite narrow. Fogel's material (Chapter 9 of this volume) suggests that one of the costs of maintaining high fertility may be lower levels of nutrition and earlier mortality. Here we have seen that rural Sierra Leoneans adjust the timing or spacing of their births, rather than prevent births. Families are best seen as entrepreneurs who adopt a portfolio strategy of fertility, diversifying children's training and opportunities, and investing disproportionately in the most promising children.[8]

Why should people choose such options rather than reduce fertility? Rural Sierra Leoneans have long lived in fear that their crops will fail or that they will become sick and be unable to perform vital physical work. Now they live in fear as well that their government and their precarious patronage network will collapse; that their currency will be declared worthless overnight; that petrol will disappear in the country and, with it, the fragile transportation infrastructure on which they now depend. Even household members cannot be trusted. There is often implicit competition among co-wives and even between spouses, whose interests may conflict and whose future paths may diverge. (For a general discussion of Africa, see Guyer 1981; for the developing world, see Folbre 1986.) Under such conditions, people try to achieve security through the domain of life that offers both diversity and stability: children. Paradoxically, children offer the greatest hope of stability because they do *not* remain the same, but can change in response to new situations. While we may imagine tradition to be the cultural core, and diversity and change to be conditions of aberration, the opposite is true. Diversity and adaptability are the heart of survival. (For similar conclusions from ecological and economic points of view, respectively, see Sen, Chapter 3, and Holling, Chapter 4 of this volume.)

Reproduction, of course, remains a process rife with uncertainty. Although children should yield access to worlds of value and security for their families through powerful patronage networks, a new-born is a creature of unknown potential. He may now prove 'clever', hard-working, or dutiful. Since children's potentials unfold over time, parents and sponsors continually reassess children and foster diversity among them. They invest large amounts in children who begin to show promise of success, waiting for them to help 'develop' younger children and edging away from obligations to those children headed for failure.

Because children have uncertain futures, adults treat a birth not as an event that guarantees them security once and for all, but as the beginning of a long, continuously negotiated relationship: a process that may conflict with other people's efforts to make claims on the children or with the children's own desires to forge more independent lives. A highly flexible and dynamic view of reproduction emerges: people constantly try to shape children and their relationships to them after the relatively unimportant event of a birth. Children and their potentials are carefully cultivated, not allowed to grow formlessly like wild bamboo trees, at the whim of nature and to the detriment of all members.

Although biological fertility provides the raw material for security, its capacity to respond to families' changing needs is severely circumscribed. Fertility limitation comprises a small wedge of the available

corpus of social, geographical, and temporal options for managing reproduction.[9] Rather than throwing up our hands in exasperation at this
apparent demographic mess, we should take it as a pivotal point of departure for studying fertility in Africa. Reproductive regimes that spread the
costs of children over time, space, and rank reveal a very broad spectrum
of strategies (some more, some less desirable) that can be employed
before invoking measures such as celibacy, infanticide, or birth control
that are commonly recorded in other times and places—options that cut
off reproductive options more sharply.

While Western parents seem to make fertility decisions based on a
minimalist strategy—the number of children one bears cannot exceed the
number who can be supported at the lowest ebb of family fortunes—most
adults in Sierra Leone view efforts to reduce fertility as cutting their
diversity options as well as their chances of producing that rare, highly
successful child. The Sierra Leone case suggests that more precarious
economic circumstances will hardly generate enthusiasm for fertility
reduction, no matter how hard agencies promote family planning. People
may instead intensify the extent to which a family differentiates the
opportunities available to its members.

Wealth alone, desirable as it is, is not what people most need in order
to take reproductive risks, for wealth can abruptly come and go. What
people most need, as Chayanov implied, is predictability. In its absence,
producing a large, diverse group of children remains families' best hope
of flattening the steepest peaks and troughs in a fundamentally insecure
world.

NOTES

1. Greenhouse gas emission provides an apt example of resource imbalance;
   according to Bongaarts (1992), because of rapid population increases, the
   total amount of CO_2 emission by 'less developed countries' (LDCs), now
   lower than that of 'more developed countries' (MDCs), will surpass that of
   MDCs by the year 2100 (14.2 versus 12.0 petrograms of carbon/year). Reducing population growth in the LDCs, therefore, could produce significant
   savings in the emission of greenhouse gases. However, imbalances, both present and future, give pause. While the MDCs had far fewer people in 1985
   than did LDCs (1.2 versus 3.7 billion), their total emission exceeded that of
   the LDCs (3.9 versus 2.2 petrograms of carbon/year), and their per-capita rate

of CO_2 emission was five times greater than that in the LDCs (3.1 versus 0.6 petrograms of carbon/year), a ratio that will probably remain stable through the next century. And, although the LDCs' total CO_2 emission will exceed that in MDCs in 2100, their per-capita emission at that time (1.6) will still fall far short of MDCs' current levels (3.1). Moreover, while both MDCs and LDCs will almost triple their emission levels of CO_2 in the next century, most of the CO_2 increases will stem from growth in GDP—i.e. development—not from population increase *per se*.

2. Fosterage has drawn recent attention in demography because of its potential explanatory value for high fertility in Africa. Building on works such as Caldwell's 'wealth-flows' argument (1982), Dasgupta (1993) observes that fosterage may create a 'commons' mentality for parents who have high fertility because they can free-ride on other people's contributions to their children's costs. I found as well that the fact that fosterage can help buffer the costs of children, especially during seasonal hardships or when several children are born in rapid succession, renders parents' biological fertility to a certain extent independent of economic costs (Bledsoe 1990; see also Caldwell 1977; Oppong and Bleek 1982; Page 1989). Indeed, in my 1985 survey, high fertility was closely associated with whether a woman was fostering out any children. (Conversely, fostering in children was closely associated with low fertility.) Fosterage, moreover, is only the tip of the iceberg of practices involving cost-sharing. Relatives of all kinds can be asked to help with the expenses of children, a pattern that is common across the continent even where fostering is not. Tracing children's costs, and how they are paid, remains a complex task. Parents who receive help for their children's costs feel bound to reciprocate in one form or another, and may end up paying as much over time as if they had paid only their own children's costs. Furthermore, although high fertility is associated with high fosterage rates within Mende areas, the two variables are linked more tenuously at the macro-level. The Mende have very high rates of fosterage even for Africa, but their completed family size (about 5.7 children in surveys from two different rural areas) was far lower than in parts of East Africa, where fosterage rates are low (see Page 1989, for figures from seven World Fertility Survey countries). Clearly the whole issue continues to pose an enormous puzzle.

3. Gudrun Dahl (personal communication) reports strikingly similar patterns for East African pastoralists: people try to maintain many different kinds of animals. They also try to purvey profits from successful pastoralism into further diversities: agriculture, trade, and schooling for children.

4. This observation suggests that, although we usually consider the wives and children currently living in a household to be the result of decisions taken at the outset of conjugal life, the family that we actually see before us may have

resulted from a process of *ex post facto* shaping. A man may have married several times but, over time, have shed other women and children whose utility was ebbing—either through divorce or by reconstruing the women as 'girlfriends' or 'country wives' and their children as 'outside' or 'illegitimate' children.

5. In rural areas of coastal West Africa, villagers used fosterage like marriage alliances, to create ties with the coastal élite. But the phenomenon of children living in non-natal households, coupled with stark inequalities manifested in institutions such as slavery and clientship, has a deep history. Oral accounts reveal that, before the twentieth century, raiding parties in the interior might kill or enslave the men in a village, taking the women as wives and the children as slaves or wards. To protect their families during these dangerous times, adults often sent children to chiefs with powerful war medicines and large bands of warriors. Boys would become bodyguards and servants within the household of the chief, who supervised their training as warriors and required them eventually to take up arms. Girls might be trained in the chief's household by his wives and married to him when they matured. But aside from these dramatic instances, children were shifted around quite casually. A mother might leave her toddler with an older girl or an old woman while she worked on the farm during the day or slept in her husband's house at night. A man might send his wife's first child to her parents as a token of appreciation. A young wife could bring in a younger sister for help with housework and farming or a younger brother for support in case of trouble with her husband or in-laws. And a chief might send a child from his village to study with a highly reputed carver, hunter, or blacksmith so that he could then return to work for the chief as a skilled tradesman.

6. Although the number of contraceptive users is quite small in our study population—about 9 % of women were using any method; 5 % was using Western ones—it is a mistake to read the results in static terms, as a tiny group of innovative 'acceptors'. Women only practice contraception for very short periods of time, after their menstrual periods have resumed, in order to prolong the birth interval just a few months. Since a very small number of women are actually using contraceptives at any given point in time, the identity of the users is shifting rapidly. If we counted over a two-year interval we would doubtless find many users eagerly exploiting the potentials of contraceptives. For more details, see Bledsoe *et al.* (1994); see Page and Lesthaeghe (1981) on the general issue of child-spacing in Africa.

7. Developing a parallel argument, Willis (1987) suggests that what parents most need in order to reduce their births is confidence that they can obtain market-rate credit and old-age support from children.

8. I am indebted to Tommy Bengtsson for pointing out the link to Chayanov.

9. Dasgupta's (1993) observations about fertility free-ridership in areas like Africa could be extended to argue that the phenomenon is a global one, with wealthy Western countries free-riding on the reproductive investments of many developing countries. In Africa, for example, families continue to need high fertility, to sustain a bare level of subsistence in their impoverished economies and to produce a few high-achieving members who can be deployed to seek resources from the cosmopolitan world. But as much as high fertility in Africa alarms the West, the West probably benefits immensely from it. In the case of the United States, the more tightly the gates are squeezed shut by the Immigration and Naturalization Service, the more high fertility in Africa benefits Americans. While less successful African children remain behind to maintain family subsistence and to underwrite investments for new cohorts of promising children, only the most educated, successful children are allowed to slip through the crack to bring in their skills and bank accounts. The United States is then spared not only the costs of raising and educating them but also (more importantly) the costs of raising the multitudes of children from whom the tiny selections were drawn.

The United Nations and the World Bank are extreme cases of US-based institutions that reap the advantages of highly educated people from developing countries. Universities are also examples. The headlines of Northwestern University's recent College of Arts and Sciences newsletter alluded to this phenomenon: 'World-class faculty: foreign-born scholars enrich the College.' Among faculty members in the college, boasted the article, 22% are foreign-born, a percentage that appears to be growing; among assistant professors hired in the last six years, 32% were foreign-born (*Mosaic: News from the College*, Fall, 1993). Such numbers hint at an unwritten pecking order: the more foreign-born faculty, the more prestige. According to this measure, Harvard and Chicago clearly have more prestige than my university; Iowa State less.

REFERENCES

Appadurai, A. (1986), 'Introduction: Commodities and the Politics of Value', in Appadurai (ed.)*The Social Life of Things: Commodities in Cultural Perspective* (Cambridge), 3–63.

Birdsall, N., Fei, J., Kuznets, S., Ranis, G., and Schultz, T. P. (1979), 'Demography and Development in the 1980s', in P. M. Hauser (ed.), *World Population and Development: Challenge and Prospects* (Syracuse, NY).

Bledsoe, C. (1990), 'The Politics of Children: Fosterage and the Social Management of Fertility Among the Mende of Sierra Leone', in W. P. Handwerker (ed.), *Births and Power: the Politics of Reproduction* (Boulder, Colo.), 81–100.

Bledsoe, C. (1991), 'The Trickle-Down Model within Households: Foster Children and the Phenomenon of Scrounging', in J. Cleland and A. G. Hill (eds.), *The Health Transition: Methods and Measures* Health Transition Series no. 3 (Canberra), 115–31.

Bledsoe, C. H., and Brandon, A. (1992), 'Child Fosterage and Child Mortality in Sub-Saharan Africa: Some Preliminary Questions and Answers', in E. van de Walle, G. Pison, and M. Sala-Diakanda (eds.), *Mortality and Society in Sub-Saharan Africa* (Oxford), 279–302.

Bledsoe, C., and Cohen, B. (1993) (eds.), *The Social Dynamics of Adolescent Fertility in Sub-Saharan Africa*, (Working Group on the Social Dynamics of Adolescent Fertility in Sub-Saharan Africa, National Research Council; Washington, DC).

Bledsoe, C., Hill, A. G., D'Allessandro, U., and Langerock, P. (1994), 'Constructing Natural Fertility: The Use of Western Contraceptive Technologies in Rural Gambia', *Population and Development Review*, 20: 1.

Bleek, W. (1987), 'Family and Family Planning in Southern Ghana', in C. Oppong (ed.), *Sex Roles, Population and Development in West Africa: Policy-Related Studies on Work and Demographic Issues* (Portsmouth, NH), 138–53.

Bongaarts, J. (1992), 'Population Growth and Global Warming', *Population and Development Review*, 81: 299–319.

Boswell, J. (1987), *The Kindness of Strangers: The Abandonment of Children in Western Europe from Late Antiquity to the Renaissance* (New York).

Boulding, K. E. (1959), 'Foreword', in T. R. Malthus, *Population: The First Essay* (Ann Arbor, Mich.), pp. v–xiv.

Brass, W., and Jolly, C. L. (1993) (eds.), *Population Dynamics of Kenya* (Working Group on Kenya, National Research Council; Washington, DC).

Brooks, C. C. (1976), 'Adoption on Manihi Atoll, Tuamotu Archipelago', in I. Brady (ed.), *Transactions in Kinship: Adoption and Fosterage in Oceania* (ASAO Monograph no. 4; Honolulu), 51–63.

Caldwell, J. C. (1977), 'The Economic Rationality of High Fertility: An Investigation Illustrated with Nigerian Survey Data', *Population Studies*, 31: 5-27.

Caldwell, J. C. (1982), *Theory of Fertility Decline* (London).

Caldwell, J. C., and Caldwell, P. (1987), 'The Cultural Context of High Fertility in Sub-Saharan Africa', *Population and Development Review*, 13: 409–37.

Caldwell, J. C., and Caldwell, P. (forthcoming), 'Marital Status and Abortion in Africa', in C. Bledsoe and G. Pison (eds.), *Nuptiality in Sub-Saharan Africa: Current Changes and Impact on Fertility* (Oxford).

Carroll, V. (1970), 'Introduction: What does Adoption Mean?', *Adoption in Eastern Oceania* (ASAO Monograph no. 1; Honolulu).

Chayanov, A. V. (1986), *The Theory of Peasant Economy* (Madison, Wis.).

Comaroff, J., and Roberts, S. (1981), *Rules and Processes: The Cultural Logic of Dispute in an African Context* (Chicago).

Dasgupta, P. (1993), 'The Population Problem', paper presented at the Conference of the World's Scientific Academies on World Population, New Delhi, 24–7 October.

Eloundou, P. E. (1993), 'Why Trade Quantity for Child Quality? A Family Mobility Thesis', Department of Sociology, Pennsylvania State University.

Fischer, A. (1963), 'Reproduction in Truk', *Ethnology*, 2: 526–40.

Folbre, N. (1986), 'Hearts and Spades: Paradigms of Household Economics', *World Development*, 14: 245–55.

Frank, O., and McNicoll, G. (1987), 'Fertility and Population Policy in Kenya', *Population and Development Review*, 13: 209–43.

Goody, E. N. (1982), *Parenthood and Social Reproduction: Fostering and Occupational Roles in West Africa* (Cambridge).

Greenhalgh, S. (1994), 'Controlling Births and Bodies in Village China', *American Ethnologist,* 21: 3-30.

Guyer, J. I. (1981), 'Household and Community in African Studies', *African Studies Review*, 24: 87–137.

Guyer, J. I. (forthcoming), 'Lineal Identities and Lateral Networks: The Logic of Polyandrous Motherhood', in C. Bledsoe and G. Pison (eds.), *Nuptiality in Sub-Saharan Africa: Current Changes and Impact on Fertility* (Oxford).

Guyer, J. I. (1991), 'The Economics of Adolescent Fertility', background paper for C. Bledsoe, and B. Cohen (eds.), *The Social Dynamics of Adolescent Fertility in Sub Saharan Africa* (National Research Council; Washington DC, 1993).

Hajnal, J. (1982), 'Two Kinds of Preindustrial Household Formation System', *Population and Development Review*, 8: 449–94.

Isiugo-Abanihe, U. C. (1985), 'Child Fosterage in West Africa', *Population and Development Review*, 11: 53-73.

Keesing, R. M. (1970), 'Kwaio Fosterage', *American Anthropologist*, 72: 991–1020.

Knodel, J., Chamratrihiron, A., and Debavalya, N. (1987), *Thailand's Reproductive Revolution: Rapid Fertility Decline in a Third-World Setting* (Madison, Wis.).

Kopytoff, I. (1986), 'The Cultural Biography of Things: Commoditization as Process', in A. Appadurai (ed.), *The Social Life of Things: Commodities in Cultural Perspective* (Cambridge), 64–91.

Lesthaeghe, R. (1989) (ed.), *Reproduction and Social Organization in Sub-Saharan Africa* (Berkeley, Calif.).

McCracken, G. (1983), 'The Exchange of Children in Tudor England: An Anthropological Phenomenon in Historical Context', *Journal of Family History*, 8:

303–13.

Malthus, T. R. (1803), *An Essay on the Principle of Population*, 2nd edn., ed. P. James, incorporating changes and additions of 1806, 1807, 1817, and 1826 (Cambridge, 1989).

Marguerat, Y. (1990), 'Les "Smallvi" ne sont pas des "Gbevouvi": Éléments pour une histoire de la marginalité juvenile à Lomé (Togo)', Paper presented at the International Conference 'La Jeunesse en Afrique dans la Societé Contemporaine Heritages, Mutation, Avenir', Paris.

Miller, B. D. (1987), 'Female Infanticide and Child Neglect in Rural North India', in N. Scheper-Hughes (ed.), *Child Survival: Anthropological Perspectives on the Treatment and Maltreatment of Children* (Dordrecht), 135–44.

Oppong, C. (1973), *Growing Up in Dagbon* (Accra).

Oppong, C., and Bleek, W. (1982), 'Economic Models and Having Children: Some Evidence from Kwahu, Ghana', *Africa*, 52: 15–33.

Page, H. J. (1989), 'Childrearing versus Childbearing: Coresidence of Mother and Child in Sub-Saharan Africa', in Lesthaeghe (1989), 401–41.

Page, H. J., and Lesthaeghe, R. (1981), *Child-Spacing in Tropical Africa* (New York).

Richards, P. (1985), *Indigenous Agricultural Revolution: Ecology and Food Production in West Africa* (London).

Scrimshaw, S. (1978), 'Infant Mortality and Behavior in the Regulation of Family Size, *Population and Development Review*, 4: 383-403.

Sen, A. (1981), *Poverty and Famines: An Essay on Entitlement and Deprivation* (Oxford).

Skinner, G. W. (1993), 'Conjugal Power in Tokugawa Japanese Families: A Matter of Life or Death', in B. D. Miller (ed.), *Sex and Gender Hierarchies* (Cambridge), 236–70.

van de Walle, E., and Ebigbola, J. A.(1987) (eds.), *The Cultural Roots of African Fertility Regimes:* Proceedings of the Ife Conference, 25 February–1 March 1987, Department of Demography and Social Statistics, Obafemi Awolowo University, and Population Studies Center, University of Pennsylvania.

von Tunzelman, G. N. (1986), 'Malthus' "Total Population System": A Dynamic Reinterpretation', in D. Coleman and R. Schofield (eds.), *The State of Population Theory: Forward from Malthus* (Oxford), 65–95.

Willis, R. J. (1987), 'Externalities and Population', in D. G. Johnson and R. D. Lee (eds.), *Population Growth and Economic Development: Issues and Evidence* (Working Group on Population Growth and Economic Development, National Research Council; Madison, Wis.), 661–702.

Winch, D. (1987), *Malthus* (Oxford).

Wrigley, E. A. (1983), 'Malthus's Model of a Pre-Industrial Economy', in J. du Pacquier (ed.), *Malthus Past and Present* (London), 111–24.

6

Economic Analysis of Fertility: Micro Foundations and Aggregate Implications

Robert J. Willis[*]

Perhaps the greatest challenge facing students of population is to understand the causes of the historical association between economic growth and development and the demographic transition from high to low levels of fertility and mortality. Every 'advanced country' that has successfully achieved a high level of living has undergone such a demographic transition. Moreover, a substantial fertility decline is currently under way in a number of countries now undergoing rapid economic growth in East and South-East Asia and, to a lesser extent, in Latin America. Conversely, it is much less apparent that countries in South Asia and, especially, in Africa are as yet on a path either of sustained economic development or of demographic transition.[1]

In addition to purely scientific reasons, it is of obvious importance to seek an understanding of the connection between economic development and demographic transitions in order to develop policies that serve broadly accepted goals of increasing the well-being of current and future generations of human beings in all parts of the world, including protecting the natural and biological resources that contribute to their well-being.[2] In this chapter, I will argue that modern developments in the economic analysis of fertility behaviour can help us understand some of these issues.

The broad thesis of the chapter is that, in order to advance our understanding of the causes and consequences of demographic change and their implications for policy, we need to integrate theories of aggregate demographic–economic interaction, such as Malthusian theory and its modern successors, and micro-level theories of family decisions about fertility, investment in human capital, and allocation of time that have

* Professor of Education and Public Policy, University of Chicago.

been developed during the past three decades. In this chapter I discuss both the implications of aggregate-level models and issues of micro-foundations. On the basis of these theories I advance a broad hypothesis to explain why the association between economic growth and demographic transition is such a strong empirical regularity. Put simply, the hypothesis is that Engel's Law—that is, the low-income elasticity of demand for food—together with the skill-using bias of technological change generate increases in the optimal level of human capital investment per child and increases in the relative cost of female time that cause desired fertility to decline and the average quality of workers to increase, which, in turn, tends to foster economic growth and development. I also discuss why the organization of the family is critical for the realization of such 'virtuous interactions', and provide some examples of situations in which negative interactions between social and economic change and demographic behaviour may occur.

1. AGGREGATE POPULATION THEORY AND POPULATION POLICY

It is often argued that the empirical regularity represented by the demographic transition provides a decisive refutation of the Malthusian theory of population. Indeed, the failure of the Malthusian model to account for observed demographic behaviour is commonly given as the reason why population dropped out of economic theory in the late nineteenth century.[3] Population did not reappear as an economic variable until the emergence of interest in theories of economic growth and development after the Second World War, and then only as an exogenous variable which grew for reasons determined outside the economic system.

Ironically, despite the rejection of Malthusian theory as a descriptive model of the relationship between demographic and economic change, much thinking about population policy during the 1950s and 1960s—and even today—is based on the Malthusian model or its analytical child, the neo-classical theory of economic growth. In the Malthusian model, diminishing returns to finite resources inexorably causes a reduction of per-capita consumption as the size of the population increases. Conversely, the model suggests that fewer people could be made happier if the size of the population is reduced. Neo-Malthusian arguments for anti-natalist family-planning policies usually turn on their proponents' preference for a high standard of living for the few, whereas anti-Malthusian arguments by the Catholic Church or the early Mao argue for the value of higher population size for moral or political reasons even at the sacrifice of the economic welfare of the average person. It is notable that neither side in these arguments finds it necessary to consider the preferences of the people whose actions determine the size of the population, except to the extent that individual preferences (and consequent behaviour) constitute an obstacle to the achievement of the demographic outcomes desired by the policy proponent.

The gloomy prophecies of Malthus were replaced by a far more optimistic vision of long-run growth in the living standards of the common man with the development of neo-classical growth theory in the 1950s. The pioneering and prototypical Solow (1956) growth model removed the barrier of finite resources by arbitrary assumption and added the blessing of exogenous technological progress, an assumption that appeared to be justified by the history of continued progress since Malthus's day but is perhaps less certain today after the 'productivity slow-down' of the past two decades. In combination, these assumptions imply that per-capita consumption may grow without bound even as the size of the population itself grows without bound. Despite this extraordinarily—and incredibly—optimistic vision, the implications of the neo-classical growth model for population policy proved to be the same as those of the Malthusian model and for essentially the same reasons, except that the population's growth rate rather than its size becomes the key variable. In place of a fixed stock of natural resources, the neo-classical growth model assumes that the productivity of each worker depends on the amount of capital with which he may work, and that the marginal product of labour diminishes as the labour–capital ratio increases. Although diminishing returns to labour caused by population growth can be offset by augmenting the stock of capital through investment, this requires diversion of output away from consumption. It follows that per-capita consumption can be increased by reducing the rate of population growth, because the lower the rate of growth, the less output needs to be allocated to investment in order to equip each worker with a given amount of capital. I shall refer to this as the 'capital dilution effect' of population growth.

An elaborated version of the Solow model was used in an enormously influential study by Coale and Hoover (1958) in order to provide quantitative measures of the impact of alternative demographic scenarios for the development plans of a low-income country. Using India as a model, their work showed that a programme of fertility reduction would considerably ease the savings and investment requirements needed to achieve any given target rate of growth of income and per-capita living standards. Results of this type, which stem entirely from the capital dilution effect described above, provided the basis for many 'cost–benefit' calculations in evaluations of family-planning programmes during the 1960s and 1970s.

The logical basis of population policy based on such calculations was called into question by Blandy (1974). Assume that an increase in population growth will reduce per-capita income and consumption because of Malthusian diminishing returns or because of capital dilution. One might think that such an effect represents a 'reproductive externality' in the sense that an additional child added to the population will reduce the average and marginal products of all workers in the future. That is, in deciding how many children to bear, an individual household has no incentive to count in the loss in wages and consumption that their additional child will inflict on the children of others. Thus, it might be argued, privately optimal decisions by parents may produce too many children from a social point of view.

Blandy suggests that this argument is flawed because, making a distinction introduced into economics by Pigou in the 1920s, the reproductive externality described above is not a 'technical externality' which leads to inefficient outcomes. Rather, it is a 'pecuniary externality', which may influence the distribution of income but has no adverse effect on economic efficiency. To illustrate Blandy's argument in the simplest setting, consider a situation in which we evaluate the impact on economic welfare of increasing a nation's labour supply by allowing some additional voluntary immigrants.[4] Because of diminishing returns, the migrants cause the wage rate to fall and also reduce the per-capita income of the *ex post* population of natives plus immigrants relative to the *ex ante* population of natives. Who are the winners and losers? Clearly, the immigrants reveal that they gain by voting with their feet. The effects of the immigration on the natives is mixed. The labour incomes of native workers fall, but the incomes of the owners of capital and land increase. In fact, the model implies that the total income of the natives (that is, the sum of their incomes from labour, capital, and land) must increase. This implies that a coalition of the owners of capital and land—the winners—could compensate the native workers for their income losses and still have something left over. If the winners could solve the collective-action problem associated with forming such a coalition, a policy to allow increased immigration is a Pareto improvement and could command a unanimous vote. If they cannot, the welfare effects of the policy remain ambiguous. Moreover, it could easily prove to be a political loser, since the potential immigrants cannot vote and the owners of non-human resources may be outnumbered.

Obviously, analyses of the impact of fertility and immigration policies may diverge because children do not choose to be born.[5] However, this difference does not affect the analysis of the *external* effects on you or your children's welfare of an addition to the size of the population caused by a decision by my wife and me to have an additional child. In particular, it would not be possible for a coalition of those who lose as a result of our child's birth to compensate those who gain and still have something left over with which to bribe my wife and me not to have the additional child.

2. AGE STRUCTURE AND INTER-GENERATIONAL TRANSFERS

The introduction of 'real' demography into economic models in the form of an age-structured population brought out the possibility that population growth could generate beneficial 'inter-generational transfer effects' that may offset the negative impact of population growth on per-capita economic welfare that is predicted by both the Malthusian and Solow models. Samuelson's (1958) seminal overlapping-generations model introduced the basic framework, although economists did not begin to explore the demographic aspects of the model until nearly twenty years later.

In the simplest version of the overlapping-generations model, it is assumed that each individual lives for two periods, first as a productive worker and later as an unproductive retiree. Output, produced by labour alone, is a non-durable consumption good like ice-cream which must be eaten immediately or it will spoil. Even though each person would like to spread his consumption over his lifecycle, Samuelson showed that private markets for consumption loans will fail and that retirees will starve in competitive equilibrium. The reason is that a young worker who wishes to give up some of his ice-cream now in exchange for a promise to deliver ice-cream to him next period will be unable to find anyone with whom he can trade. His elderly contemporaries can do nothing for him; he can do nothing for the (unborn) workers of the next period.

Samuelson argued that some kind of social institution is needed if society is to compensate for this market failure and solve the problem of the starving elderly. He gave three examples: a family system of transfers from children to their elderly parents; a pay-as-you-go social security system in which workers are taxed and the proceeds tranferred to the elderly; or the creation of 'fiat money' as a store of value. The introduction and continuation of a system of transfers from the younger to the older generation through any of these institutions will lead to an improvement in welfare of all persons in the current and all future generations. Yet members of each generation have an incentive to renege on the arrangement. Because of this, Samuelson suggested that transfer institutions need to be supported by a 'social compact' which, by constitutional means or through internalized norms, enables their survival in the face of powerful selfish incentives which would destroy them.

Almost two decades after introducing the overlapping-generations model, Samuelson (1975) initiated what has proved to be a sustained line of research which explores the implications of population growth for economic welfare in age-structured populations. He began by pointing out that an increase in the rate of population growth generates a positive effect on potential economic welfare within an overlapping-generations model of the type under discussion because of a positive 'inter-generational transfer effect' which tends to offset the negative 'capital dilution effect' caused by increased growth within a Solow-type model. The inter-generational transfer effect arises because the more rapid the rate of population growth, the larger is the ratio of workers to retirees in the population and, hence, the lower is the transfer from each worker that is required to maintain a retiree at any given consumption level.[6] For example, a zero rate of population growth implies that there will be equal numbers of workers and retired, while, if population were doubling every generation, there would be two workers for every retiree.[7] With zero growth, a social compact requiring each worker to give up one unit of ice-cream would enable each retiree to consume one unit, while, with population doubling, the same sacrifice by workers would enable each retiree to eat two units of ice-cream. Conversely, population decline has a negative impact on per-capita welfare because it increases the fraction of dependent elderly in the population, creating a higher tax burden on those of working age.

Although Samuelson's overlapping-generations model introduced demography into economic models in a fundamental way, it did so in a highly artificial manner that does not lend itself to empirical application. Arthur and McNicoll (1978) took an important step towards making the model more realistic by considering a population with a full age structure. Like Solow, they assume that aggregate output is a function of aggregate labour and capital and that aggregate investment is equal to the difference between aggregate income and aggregate consumption. They introduce age structure by assuming that each individual has age-specific schedules of labour productivity and consumption. Aggregate quantities of labour or consumption are simply equal to the sum of age-specific labour supplies or age-specific consumption multiplied by the number of people at each age. Finally, they assume that the economy is in a steady-state golden rule equilibrium in which the rate of population growth is equal to the rate of interest.[8]

Given this set-up, Arthur and McNicoll ask whether a small increase in the rate of population growth will increase or decrease the level of per-capita welfare in steady state. The answer comes in a beautifully simple expression given by the weighted sum of a capital dilution effect which is always negative and an inter-generational transfer effect whose sign depends on the difference between two quantities that they call the 'average age of consuming' and the 'average age of producing'.[9] The intuition underlying this result is seen most easily by referring back to the simple Samuelson model with two age groups, the young and the old. Assume that consumption is spread evenly over the lifecycle. In Samuelson's original model, the young are assumed to be productive and the old unproductive, so that, on average, consumption takes place at an older age than production. In this case, the direction of transfers is from the younger to the older generation and it is advantageous to have a higher growth rate in order to increase the fraction of the population at young ages. Note, however, that the analysis is exactly reversed if the young are unproductive and the old are productive, so that the average age of consuming is less than the average age of producing. Since the old must transfer to the young if each individual is to have a smooth lifecycle consumption path, increased population growth is disadvantageous because it increases the proportion of those dependent on transfers.

3. ECONOMIC GROWTH AND CHANGES IN THE SOCIAL AND PRIVATE COSTS OF POPULATION GROWTH

The connection between the lifecycle structure of production and consumption, the direction of inter-generational transfers, and the welfare effects of population growth revealed in Samuelson's model and its extension by Arthur and McNicoll provides an important clue about the relationship between economic growth and fertility transition. In particular, because of Engel's Law, increases in per-capita income caused by productivity change throughout the economy tend to reduce the fraction of income accounted for by agricultural output and to reduce the fraction of the labour force in the agricultural sector. The strength of this empirical relationship across countries is illustrated in Fig. 6.1a by the strong negative relationship between the share of GDP generated by agriculture and per-capita income and, in Fig. 6.1b, by the negative relationship between the growth of real GDP between 1965 and 1988 and the change in the share of agriculture in GDP over the same period. Since the agricultural sector tends to employ the least-educated labour in an economy, whatever the level of development, Engel's Law implies that economic growth will tend to increase the relative demand for skilled labour, causing an associated increase in the demand for education and on-the-job training. Although I have not seen it documented, I conjecture that changes in economic structure associated with economic growth of the type documented by Kuznets (1966) generate a broader correlation between the skill-intensity of sectoral labour demand and sectoral employment growth.[10]

Assuming that it is 'skill-using' in the sense described above, economic development tends to shift the age structure of lifecycle productivity towards later ages, thereby increasing the average age of producing relative to the average age of consuming.[11] As a result, the social cost of an additional birth is increased in the sense that a small increase in the rate of population growth reduces the feasible level of lifecycle consumption

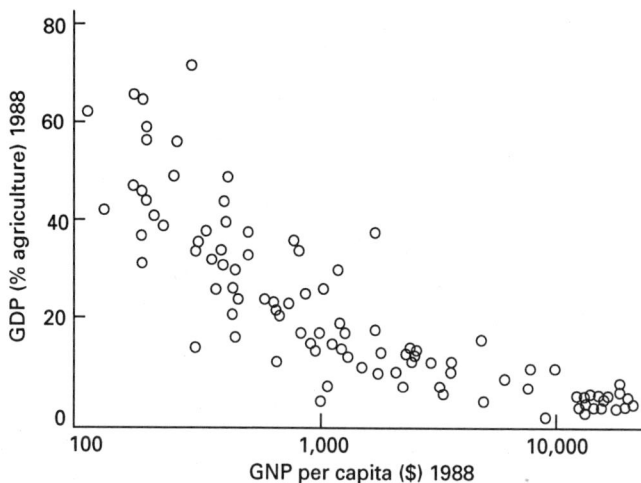

(*a*) Percentage GDP in agriculture v. per-capita income, 1988

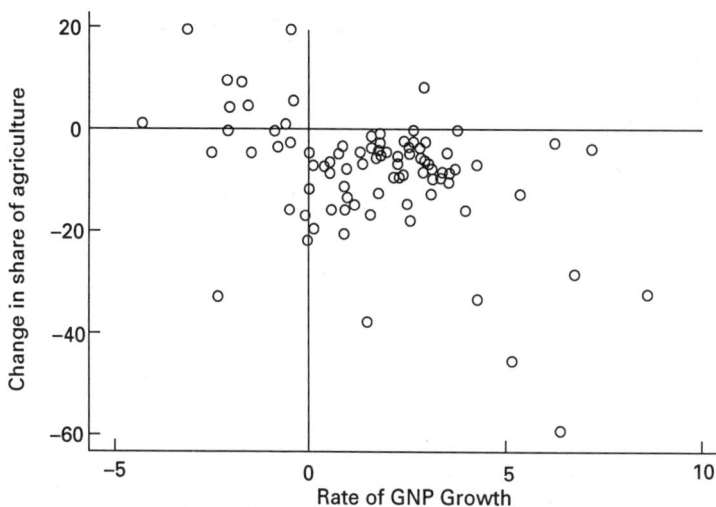

(*b*) Change in agriculture share v. income growth, 1965–1988

Source: World Bank (1992).

FIG. 6.1 Illustrations of Engel's Law

that may be enjoyed by members of each generation in steadystate (Willis 1982, 1988). Equivalently, an increase in the average age of producing increases the magnitude of net inter-generational transfers from the older to the younger generation that are required to maintain a given rate of population growth. To the extent that increases in the social costs of births are translated into increased private costs of fertility to the individuals making reproductive decisions, this creates pressure for reduced fertility.

Divergences between the private and social costs of population growth may occur when inter-generational transfers take place through the public sector. In a steady-state golden rule economy, this divergence is measured by the net public-sector transfer whose sign and magnitude is proportional to the difference in mean ages of transfer receipt and tax payment, where these mean ages are defined analogously to the mean ages of producing and consuming in the Arthur–McNicoll model (see Willis 1988).

As an obvious example, a pay-as-you-go social security programme represents a public-sector transfer from the younger to the older generation because the average age of those receiving benefits exceeds the average age of those making tax payments. As has been noted by several writers (Willis 1980; Becker 1981; Wildasin 1990), a social security programme creates a divergence between the private and social costs of child-bearing because the costs of rearing a child are borne privately by its parents while the benefits from the social security taxes that the child will pay in the future are diffused across the entire population. Citizens of advanced countries today, especially Europeans, have come to fear the consequences of population decline which increases the fraction of elderly in the population, creating a potentially intolerable tax burden on those of working age. The analysis here suggests that a society that relies on the younger generation to make net transfers to the older generation through the public sector runs the risk that the older generation will respond by engaging in 'reproductive shirking' (Willis 1980).

In contrast to the case of social security, public-sector transfers from the older to the younger generation, such as subsidized schooling, tend to make the private cost of children lower than their social cost. In many developing countries, transfers in this direction probably dominate those in the opposite direction both because public education tends to develop more quickly than public pensions during the development process, and because the age distribution in developing countries is skewed towards younger ages, primarily because of high fertility and secondarily because of high mortality.

Although analysis of aggregate accounting relationships within an age-structured population, such as that discussed in this section, can provide important insights into economic–demographic interactions, it can be seriously misleading if not accompanied by analysis of micro-level behaviour. The case of publicly subsidized education provides a good example of this point. The analysis presented in this section might suggest that public schooling is a prop for high fertility in the Third World and that wise social policy would impose school fees to help reduce excessive population growth. As I shall argue in the next section, subsidized schooling may provide an important mechanism to overcome adverse effects of capital market constraints in determining parental decisions about investment in their children's human capital. Moreover, to understand the effect of schooling subsidies on fertility behaviour, it is necessary to consider how parents respond to trade-offs between having more children and spending more on each child. Finally, it is important to consider the long-run effects of educating one generation, especially females, on that generation's subsequent reproductive decisions through its effects on the value of time, the distribution of power within the household, and in other ways that I shall discuss later.

4. MICRO-FOUNDATIONS: THE ECONOMICS OF THE FAMILY

After its banishment in the nineteenth century, demographic behaviour was not reintroduced into economic analysis until the late 1950s when Leibenstein (1957) and, more influentially, Becker (1960) presented models in which fertility decisions were incorporated into conventional models of household behaviour. From that point onward, a wide range of human behaviours traditionally considered to be outside the scope of economic analysis has been incorporated into the theory, most often by Becker. By now, economic theories of fertility behaviour are embedded in a broader economic theory of the family which integrates traditional topics of consumer behaviour such as consumption, savings, and labour-supply decisions, and decisions about traditional demographic variables such as fertility, mortality, marriage, and migration (see e.g. Becker

1981; Cigno 1991). In addition, micro-economic theories of the family include analysis of such issues as parental investments in the health, education, and other aspects of their children's human capital; household production and the division of activity between home and market production; the sexual division of labour within the family; and the role of the family in insuring members against financial and health risks and helping them to finance investments and old-age consumption.

As a point of departure, it is useful to consider a simple economic model of fertility along the lines of one I presented two decades ago (Willis 1973). That model considers a husband–wife couple who act as a single, utility-maximizing decision-maker. Their utility depends on the adult standard of living they achieve and on the number and 'quality' of children that they have. Adult commodities and satisfaction from children are not directly purchased in the market. Rather, they are produced according to a 'household production function' with inputs of market goods (for instance, food, housing, and so on) and of adult time (for instance, time spent cooking, caring for children, going to the movies, and so on). The key empirical assumptions of the model are that mothers are primarily responsible for child care and that children require relatively more of the wife's time than do activities which contribute to the adult standard of living.

There are two main implications of this model. The first, which may be called the 'female cost-of-time hypothesis', is that increases in the value of female time tend to increase the cost of children relative to the adult standard of living, causing households to substitute away from children. Increases in the value of female time, which place downward pressure on fertility, also tend to create incentives for women to conserve their scarce and increasingly valuable time by substituting purchased goods and, as well, incentives to increase their supply of market labour in order to be able to purchase these additional goods. The consequences of the increased value of time for modern life in advanced countries are amusingly and perceptively described by the Swedish economist Stefan Linder, in his aptly titled book, *The Harried Leisure Class* (1971).

There are several channels, often mutually reinforcing, through which economic growth and development tends to increase the value of female time and, hence, to increase the relative price of children. First, general economic growth increases the value of time by increasing real wages and real household income. As Willis (1973) emphasizes, the 'shadow price of time' of wives who do not participate in the labour market tends to increase as their husband's income increases. However, the value of time in a growing economy tends to increase faster in the market sector than in the household, causing women to enter the labour force where the value of their time is determined by the market wage rate. Increased labour-force attachment by women enhances their incentives to invest in market-oriented human capital, which, in turn, further increases the value of their time.

The second major hypothesis suggested by this model of fertility behaviour is the 'quality–quantity interaction' hypothesis. Becker (1960) introduced the distinction between the number and quality of children in order to help resolve the long-standing empirical puzzle of why higher-income families tended to have fewer children. Part of the answer, he suggested, is that parents tend to increase the expenditure per child rather than the number as income goes up in much the same way that higher income leads consumers to shift towards more expensive automobiles or higher priced calories without necessarily increasing the number of cars or calories. Subsequently, Willis (1973) and Becker and Lewis (1973) showed that the demand for child quality and quantity interact because the marginal cost of another child is higher the higher is expenditure per child and, conversely, the marginal cost of increasing child quality is higher the more children there are. Thus, if income has a stronger effect on the demand for quality than for quantity, then the relative marginal cost of the number of children will tend to increase as family income increases, causing a substitution effect against fertility which may be strong enough to more than offset a positive income effect in favour of children.

This model can easily be placed into an inter-generational framework that can form the micro-foundation for the age-structured aggregate models that were described above in Sections 2 and 3. To fix ideas, consider a model in which the lifecycle is divided into three stages: childhood, adulthood, and old age. Assume that reproduction takes place at the beginning of adulthood and that a child's future productivity as an adult depends on human capital investments in his health and education that are made during childhood with inputs of the child's own time, parental time, and inputs from health providers and schools in addition to direct expenditures on the child's food, clothing, and shelter. The expected net present value of these investments depends on the cost of current inputs, including the opportunity costs of forgone child labour and parental time devoted to children; on the expected wage rates of different skill grades of labour when the child reaches adulthood; on the likelihood that the child survives to adulthood; and on the rate of discount.

Within this framework, parental decisions about fertility and investment in their children's human capital depend crucially on the nature of parental preferences. One influential line of theorizing introduced by Becker (1974) and Barro (1974) hypothesizes that parents are altruistic towards their children in the sense that they care about their children's welfare. Assuming a perfect capital market in which parents can borrow and lend at a given interest rate, and disregarding non-pecuniary returns to education, risk, and other complications, Becker and Tomes (1976) show that decisions in a household headed by an altruistic 'patriarch' can be decomposed into two steps: (1) to maximize the joint wealth of the entire family line by investing in the human capital of each child up to the point at which the marginal rate of return is equal to the rate of interest; and (2) to redistribute the resulting wealth among family members so as to maximize the preferences of the patriarch. Under these assumptions, Willis (1987) shows that the sequence of privately optimal decisions made by patriarchs in each generation yields Pareto optimal levels of fertility and investments in children from the standpoint of all current and future generations within the family 'dynasty'.

Although the 'perfect-market' model of parental altruism described in the preceding paragraph makes extremely strong and unrealistic assumptions, it provides a useful point of departure for considering the behavioural, institutional, and policy-related implications of departures from these assumptions. To set the stage for exploring some of these departures, it is helpful to consider the nature of the optimal solution in the perfect market model in a little more detail. In this solution, net parental expenditures on a given child are equal to the direct and opportunity costs of all the resources devoted to his consumption and to investment in his human capital during childhood, plus or minus the value of a monetary transfer made in the next period when the parents are old and the child has reached adulthood. Since the nature of the institutional problems faced by the family depends on the direction of intrafamily transfers between elderly parents and their grown children, I consider the two possibilities in turn. In order to focus on inter-generational relations, I temporarily consider the two parents as a unitary decision-maker. Later, I shall point out some important consequences of relaxing that assumption.

The direction of inter-generational transfers within the family will be from the older to the younger generation over the entire lifecycle if parents are sufficiently altruistic and/or sufficiently wealthy relative to their children. That is, under these conditions parents invest in their children's human capital during childhood and, in addition, make a positive monetary transfer or bequest to them at the end of life.[12] From an institutional point of view, a solution of this type is relatively easy to implement within a nuclear family structure of the sort common in developed Western societies, in which grown children leave their families of origin and establish separate and largely independent households. The reason is that there is no conflict of interest between the generations within the family because the parents are happy to give and the children are happy to receive.[13] Family investments in human capital may vary across children because of individual differences in ability or because of differences in the rate of return to investments in boys or girls. If the parents have egalitarian tastes, less able children who receive small investments in human capital will be compensated with larger monetary transfers.[14] When parents make positive monetary transfers to their children, another implication of the model is that the optimal investment in each child is independent of parental income. As parents' income increases, they tend to increase monetary transfers to each child, thus increasing total expenditure per child. Desired fertility may either increase or decrease, depending on the nature of the quality–quantity interactions on the demand for children.

The institutional problems faced by the family are quite different if the direction of transfers is from grown children to their elderly parents. It is worth distinguishing two different circumstances that might lead to such a situation. One corresponds to the case in which the parents are selfish (or only weakly altruistic), as is hypothesized by Caldwell (1976) in his well-known hypothesis that demographic transition is caused by a change in the direction of 'inter-generational wealth flows'. In this case, fertility is motivated by profit. That is, parents would not choose to have children unless the present value of transfers from children, in the form of either child labour or old-age transfers, exceeds the cost of rearing children so that the net cost of children is negative. Consequently, as

Caldwell emphasizes, parents would be motivated to have as many children as possible. If this equilibrium is to be feasible, it is important for the older generation to have sufficient power *vis-à-vis* the younger generation within the family to be confident that it will be able to extract a surplus from its children. Such a balance of power is supported by ideologies and customs which cause individuals to sublimate their self-interest in favour of familistic interests. Caldwell argues that the rise of 'individualism' spawned by the spread of education and the mass media removes this support and, with it, the props for high fertility, causing a decline in fertility.

There is a fundamental problem with this story. If parents have children for purely selfish reasons, then the same forces that remove the props for high fertility also remove parental incentives to invest in their children's health and education. In reality, of course, sustained fertility declines have been accompanied by increases rather than decreases in parental investments in their children's health and education.

In contrast to Caldwell's theory, it is quite possible to have an equilibrium in which parents are strongly altruistic towards their children, yet wealth-flows from grown children to their elderly parents are observed. This pattern occurs when parents are relatively poor and the optimal level of investment in children is relatively high. In such a situation, old-age transfers may be interpreted as the repayment by the child of an implicit loan made to him by his parents in order to cover those costs of child-rearing which exceed the amount that the parents wish to transfer to the child. However, unlike Caldwell's theory, desired fertility is limited because the marginal cost of children to the parents is positive. That is, because the parents are altruistic they choose to make the present value of the transfers they receive from a child smaller than the value of the resources they devoted to rearing him. Empirical evidence on this point is sketchy, but much of it casts doubt on the hypothesis that parents make profits on their children in peasant societies.[15]

It is important to note that old-age transfers from children discussed so far do not arise because parents have inadequate means to save for their old age; given the assumption of perfect capital markets, they can earn a safe return on their savings equal to the market interest rate. The reason that parents invest so much in their children rather than simply saving for old age is that, for all investments up to the optimal level, the marginal rate of return on human capital investments exceeds the market interest rate. In the context of a rapidly growing economy, parental finance of their children's human capital facilitates a child's capacity to exploit the improved opportunities for skilled labour that a growing economy offers and the old-age transfers from children to their parents provides a means by which the parents can share in the benefits of growth. Using data from Taiwan on intra-family transfers, Lee, Parish, and Willis (1994) find evidence consistent with such sharing rules: transfers from children to parents increase the higher the child's income, holding parents' income constant, and decrease the higher the parents' income, holding the child's income constant.

There are two potential limitations on the capacity of families to serve as a source of finance for investment in children's human capital. The first is that parents' willingness to finance such investments will be reduced the higher the risk that their children will fail to provide old-age support because they cannot (for example, because the child dies, is economically unsuccessful, or is a daughter who will owe her allegiance to her husband's family when she marries) or because they choose to default on their obligations. Clearly, these risks tend to be greater for poor parents than for rich ones, because the poor hold a larger fraction of their 'retirement portfolio' in the form of obligations from their children. Consequently, poor parents would tend to invest less in each child's human capital, although they may have more children in order to hedge against these risks.

The second limitation concerns borrowing constraints that bind when the optimal level of investment in children exceeds the amount that the parents wish to save for old age. Once again, this constraint is more likely to be encountered by poor parents. To finance this level of investment, the parent might seek to borrow from an outside source of credit. Generally, this will be unfeasible unless the parents' debts can be assigned to the child. Alternatively, the parents can self-finance the investment by sacrificing their own consumption during the child-rearing phase of life. A selfish parent who does not expect to live to old age obviously would be unwilling to make any sacrifice at all. Conversely, altruistic parents who regard their child's future welfare as a substitute for their own current consumption may be willing to make such sacrifices even if they will not be present to witness the fruits of the investment. The absolute level of parental income is likely to play a significant role in determining how responsive parents may be to improvement in their children's opportunities. For instance, even highly altruistic peasants who are very close to a subsistence level of living may have little scope for sacrificing current family income by allowing their children to go to school instead of engaging in agricultural labour.

There are several important implications of the credit constraints and other limitations on the capacity of families to make optimal investments in the human capital of their children. First, to the extent that it is the poorest families that have the most difficulty in responding to increased investment opportunities created by economic growth, the model may help to explain the 'Kuznets curve', an empirical regularity in which income inequality increases during the initial phases of economic growth. Secondly, failure of marginal rates of return to be equated across households represents an inefficiency which results in lower aggregate output growth than would be possible if the same aggregate investment were allocated efficiently. By easing credit constraints for the poor, public subsidies for schooling may promote both increased equality in the income distribution and greater efficiency of investment in human capital.

Fertility reduction is one of the most important responses by families to the economic pressures and incentives created by increased demand for human capital caused by rapid economic growth. From a policy perspective, it might be thought that there is a conflict between the benefits of public subsidies for education, discussed in the preceding paragraph, and fertility reduction. On theoretical grounds, the short-run effect of school subsidies on fertility incentives is ambiguous. The subsidy lowers the cost of a child with a given amount of schooling, but creates an incentive to increase the amount of schooling. The greater the latter effect, the more likely it is that subsidized schooling will result in lower rather than higher fertility. In the long run, however, increased education of children, especially girls, will result in increases in parental education and higher costs of female time. In his survey of the literature and in original empirical work using cross-country data, Schultz (1992) finds that increased female education has contributed powerfully to the decline of both fertility and child mortality.

To this point, I have followed a tradition in family economics of assuming that family decisions are made by a 'family decision-maker' who, somehow, represents a consensus of the individuals who make up the family. In his altruism model, Becker (1974, 1981) provides a theoretical justification for this approach by showing that a family head—or 'patriarch', to use the term most suited to the inter-generational issues discussed in this paper—who makes net transfers to other family members effectively determines the allocation of family resources as if he were a dictator. A strong implication of the altruism model is that the allocation of family resources depends only on total family income and is independent of the distribution of income within the family.

In recent years a number of economists, beginning with work by Manser and Brown (1980) and McElroy and Horney (1981), have developed alternative models of the family which emphasize bargaining and strategic interaction among family members. An implication of the bargaining models is that, in contrast to the altruism model, the allocation of family resources is not independent of the distribution of income within the family. In particular, the bargaining models imply that the welfare of women and the way in which family resources are allocated depends on the degree to which wives control family resources. A growing literature has emerged which attempts to test these alternative hypotheses empirically, including a number of studies in developing countries which have been surveyed by Strauss and Thomas (1993). Empirical findings in the literature appear to support the bargaining approach. Of more substantive interest, these studies suggest that women have relatively stronger preferences for children's welfare than do men, so that, for example, anthropometric measures of children's health are higher as women's control of family resources increases (Thomas 1990) and, further, to increase in the fraction of family income that is devoted to health, education, and housing (Thomas 1993). In addition, in a study using data from Thailand, Schultz (1990) has found that increases in the woman's share of family income reduces fertility.

The view that families are made up of persons with separate interests also has important implications for understanding how coalitions between men and women formed by marriage and broken by divorce influence the allocation of resources to children. The fact that children are—from the viewpoint of the father and mother—a 'collective good' can, as emphasized by Weiss and Willis (1985), explain both the gains to marriage, and why divorce so frequently leads the father to discontinue his economic support for his children. More recently, Willis (1993) has used this framework to produce a theory explaining out-of-wedlock child-bearing and the failure of fathers to participate in the rearing of children. In both instances, the failure of fathers and mothers to co-ordinate their resource allocation within marriage leads to inefficiently low levels of expenditure on children. Although these theories have been applied chiefly to issues concerning families in the United States, the analytic issues they address would appear quite pertinent to family orga-nizations of the sort described by Bledsoe (Chapter 5 of this volume) in West Africa or to the phenomenon of child abandonment in Brazil.

5. ECONOMIC GROWTH AND FERTILITY TRANSITION: SOME ASIAN EXAMPLES

At this point it is appropriate to ask whether the theoretical argument made in this chapter is valid from an empirical point of view. It is beyond the scope of this chapter to attempt a systematic answer to this question; I simply refer the interested reader to Schultz (1992) for a recent and thorough survey of the current empirical literature on fertility behaviour in developing countries. In place of such a survey, I will briefly discuss some examples of rapid demographic and economic transitions that have taken place recently in East and South-East Asia to illustrate some of the points made in the theory.

As can be seen in Table 6.1, since the mid-1960s many of the countries in East and South-East Asia have experienced rapid economic growth, with growth rates of real per-capita income ranging from over 6 per cent per year in Taiwan (not shown), South Korea and Singapore, to about 4–5 per cent for China, Malaysia, Thailand, and Indonesia. Laggards in the region, including the Philippines, Myamar, Cambodia, Laos, and Viet-nam, have all experienced substantial political turmoil and economic disorganization.

As of the mid-1960s total fertility rates in most countries of the region were about six children per woman, with the major exception of Japan, which had long since reached an advanced state of economic develop-ment and had completed its demographic transition, and the minor exceptions of South Korea and Singapore, in which fertility rates had begun to decline in the mid-1950s. By 1988 those countries which had experienced the most rapid economic growth had generally also expe-rienced very significant fertility declines, with fertility falling to below replacement level in South Korea and Singapore, a level similar to that

TABLE 6.1 *Economic and demographic change in Asia, 1965–1988*

	Total fertility rate		Female expectation of life		Income growth per capita	GDP per capita
	1965	1988	1965	1988	1965–88	1988
East Asia						
China	6.4	2.4	57	66	5.4	330
Korea, Rep. of	4.9	1.8	58	73	6.8	3600
Japan	2	1.7	73	81	4.3	21020
South East Asia						
Lao PDR	6.1	6.6	42	51	–	180
Indonesia	5.5	3.4	45	62	4.3	440
Philippines	6.8	3.8	57	66	1.6	630
Papua New Guinea	6.2	5.2	44	55	0.5	810
Thailand	6.3	2.5	58	68	4	1000
Malaysia	6.3	3.7	60	72	4	1940
Singapore	4.7	1.9	68	77	7.2	9070
Kampuchea Dem.	6.2	–	46	–	–	–
Vietnam	–	4	–	68	–	–
South Asia						
Bangladesh	6.8	5.5	44	51	0.4	170
Nepal	6	5.8	40	51	–	180
India	6.2	4.2	44	58	1.8	340
Pakistan	7	6.6	45	55	2.5	350
Sri Lanka	4.9	2.5	64	73	3	420
Myamar	5.8	3.9	49	62	–	–

Source: World Bank (1990).

in Japan. Fertility also declined rapidly in all the countries in which the rate of per-capita income growth exceeded 4 per cent, reaching 2.4 in China, 2.5 in Thailand, and about 3.5 in Malaysia and Indonesia. Among the laggards, fertility fell substantially in the Philippines, from 6.8 to 3.8, while the situation in Indochina is less certain because of incomplete data. For comparison, Table 6.1 includes information on these variables for countries in South Asia, where in general the pace of both economic and demographic change is markedly slower.

Rapid economic and demographic change has been associated with rapid growth in family investments in children's education and virtual elimination of gender differences in education in recent cohorts in Taiwan (Parish and Willis 1993) and Malaysia (Lillard and Willis, forthcoming), while Indonesia shows rapid convergence of male and female education (Deolalikar 1993). For example, in Taiwan, of children who reached 12 years of age during the 1940s, males averaged 6.0 years of education and females only 3.8 years; for cohorts reaching age 12 during the 1970s or later, males and females each had about eleven years of education. A similar pattern holds for Malaysia.

The economic theories of fertility described earlier in this chapter suggest that the changing pattern of labour demand during rapid economic development creates incentives for families to increase investments in their children's education and, especially for low-income families who are most likely to be credit-constrained, to reduce fertility. Parish and Willis (1993) find indirect evidence for the importance of credit constraints in family decisions about education in Taiwan. Specifically, they find that the education of a given child is sensitive to the number and gender of older and younger siblings among poorer families, while these factors become much less important among wealthier families. This suggests that low-income parents must carefully weigh how to invest in each child, taking into account the impact of these investments on their old-age security and on the education of other siblings. In constrast, wealthy parents may make the optimal investment in each child without worrying about the impact of this investment on the other children.

It is difficult to determine the causal connections among variables with as much joint determination and mutual feedback as education, income, and fertility. In an interesting attempt to get at this issue using focus group methods, Knodel, Havanon, and Pramualratana (1984) asked rural villagers from Thailand about their reasons for limiting fertility. Economic factors figured prominently in their answers and, among these factors, the trade-off between fertility and educating children seems especially important. For example, a young woman said: 'We can afford our children's education if we've got just a few. But if we had more it would be a big burden' (p. 306). Other quotations reflect the connection in people's minds between the education of their children and the kinds of jobs they will be doing when they grow up, as, for instance, in the following statement by a young woman: 'We are in difficult times as farmers. I don't want my children to do this type of work like me. I want my children to have knowledge, to do work sitting in a chair like other people' (p. 306). Finally, an older woman reveals the connection between investment in education of children and old-age security when she states: 'Children who have an education sometimes make their parents comfortable, but those without an education will depend on their parents. Those with an education will use their education to earn a living' (p. 307).

The connection between the cost of children's education and fertility may be revealed by a quasi-natural experiment in Malaysia associated with its New Economic Policy (NEP), a radical system of race-based education and employment preferences begun in 1970. This possibility is suggested in a fascinating paper by Jones (1990) that was motivated by the puzzling discrepancy between the fertility trends of Malays, the majority ethnic group in Malaysia, and trends among the Chinese and Indian minority groups. In particular, Jones noticed that the fertility of these minorities in Malaysia fell steadily from about seven births per woman in 1958 to about three in 1988, a pattern similar to fertility declines elsewhere in the region. In contrast, among the majority Malays the total fertility rate fell to about 4.5 from 1966 to 1976, but then began a mild increase which lasted through the next decade. Intrigued by this discrepancy, Jones investigated fertility patterns of Malay ethnic groups in Indonesia and Singapore and found that they had experienced fertility declines similar to individuals of other ethnicities in their home countries and similar to the declines of the Chinese and Indian ethnic groups in Malaysia. Unlike Malays elsewhere, Malays in Malaysia were beneficiaries of a government policy which provided monetary subsidies for education, required lower test scores for promotion, and provided the assurance of 'good jobs' at the end of the process to the children of Malays. With less need to sacrifice in order to improve their children's chances, it appears that Malay parents found less need to reduce their fertility. In the long run, however, the growth in the level of female education, a small part of which resulted from the NEP policy, is likely to place downward pressure on fertility among Malays.[16] The potential is considerable, since the average education of 18-year-old Malay women in 1988, who will be the mothers of the next generation, is about eleven years, while the education of their mothers averaged only about three years.

6. CONCLUSION

In this chapter I have argued that modern developments in economic demography can go some distance in explaining why economic growth and development is so strongly linked to demographic transitions from high to low levels of fertility and mortality. The crux of the argument is that economic growth leads to changes in economic structure, most notably a reduction in the share of agriculture, and corresponding increases in the share of resources going to industry and, as income growth continues, shifts from industry to the service sector. These structural shifts are associated with a radical change in the skill composition of labour demand which decreases the demand for low-skilled agricultural labour and increases the demand for relatively more-skilled labour. The supply of labour is determined in the short run by household time-allocation decisions in which men and, most crucially, women decide how much labour to devote to non-market activities within the household and how much to devote to market labour. In the intermediate run, family migration decisions determine where this labour is supplied, and on-the-job training provided by firms can begin the process of reshaping the skills of the labour force to meet the changing pattern of demand. While market forces provide powerful incentives to permit the restructuring of the labour force in an efficient and productive manner, restructuring often generates new problems associated with inadequate infrastructure and an inability to utilize natural resources effectively, sometimes resulting in vast and polluted urban slums. Moreover, public policy, which ideally is 'part of the solution' is too often 'part of the problem', creating barriers to efficient restructuring and failing to deal with the externalities it causes. Stagnation rather than growth may be the consequence.

In the long run, the size and skill composition of the labour force is determined primarily by family decisions of men and women in one generation about the number and quality of persons in the next generation. In a growing economy, changes in the structure of labour demand—indeed, expectations of such changes in the future—create incentives for parents to increase the education of their children and to reduce fertility. These incentives depend, in part, on the degree to which parents are altruistic towards their children and see investment in their human capital as a means of improving their children's welfare and, in part, on the degree to which the parents themselves can share in benefits of the increases in their children's income through inter-generational transfers from the children. A strong family system with effective bonds of trust that extend across the generations and over long periods of time greatly facilitates these responses by giving the older generation confidence that the children will not renege on their obligations if the parents forgo alternative forms of old-age security in order to finance increased investments.

From the point of view of the aggregate economy, an increase in the level of human capital investment per person causes the mean age of producing to increase relative to the mean age of consuming. Holding the rate of population growth constant, this change in the age structure of production requires an increase in inter-generational transfers from the older to the younger generation. Equivalently, this change implies that the social cost of population growth has increased. In the absence of distortions, this translates into private incentives for fertility reduction. By easing borrowing constraints, the development of publicly subsidized schooling increases the likelihood that investment in the human capital of poor children will increase in response to the opportunities contributed by economic growth. The reduction in the private cost of children produced by schooling subsidies tends to be offset, eventually, by the shift of old-age support from the family to the state with the development of social security systems. As fertility declines, causing the age structure to become older, as the length of life is increased by reduced mortality, and as the age of retirement falls, the younger generation tends to make net transfers to the older generation through the public sector, and the private costs of children tend to exceed the social costs.

What are the policy implications of theories arising from modern economic demography? First, I think that it is clear both in the history of the West and in the recent history of East and South-East Asia that transitions in fertility and mortality tend to take place in the context of rapid economic growth, even in the absence of strong anti-natalist government policies. The combination of reduced fertility and mortality, increased investment in children's human capital, and general productivity improvements and structural changes that make up the economic-demographic transition that we call economic development holds the promise of vastly improving the lives of poor people thoughout the world. Economic development holds particular promise for women. Indeed, Schultz (1992) finds that, after adjustments for endogeneity are made, family-planning programmes appear to have had a larger independent effect on women's health than on their fertility. In addition, economic development, at least in Asia, appears to reduce and even eliminate disparities in male and female education. This clearly benefits the current generation of women, both by expanding their opportunities in the labour market and by increasing their bargaining strength at home. In addition, improvements in female education will tend to reduce fertility because of increases in the cost of female time and because of quality–quantity interactions in the demand for children. Recent research also suggests that, to the extent that women's bargaining power within marriage is improved, families tend to allocate more of their resources towards children.

These considerations suggest that the Malthusian policy prescription should be reversed. Malthus argued that policies designed to improve the economic lot of the common man, such as the acquisition of new land which raises the marginal product of labour or the redistribution of income from rich landlords to poor members of the working class, will ultimately fail because they will stimulate earlier marriage and higher fertility, which, in turn, will increase the size of the population and the marginal product of labour until the worker's level of living falls back to its initial level. This is, of course, the famous 'Iron Law of Wages'. According to Malthus, the only way permanently to improve the lot of the common man is to change his reproductive behaviour. In effect, Malthus advocated that the common man should increase the level of living he requires in order to be willing to reproduce himself. The historical record since Malthus shows that the major effect of economic progress itself is to increase this required level of living.

I have said nothing in this chapter about a second area of policy concern, the interaction of population and the environment. In large measure, I have left this issue out because modern economic demography has largely failed to address such interactions. Given my own lack of expertise in this area, I will offer only a few brief comments on the policy implications of population–environment interactions.

As is true for any other aspect of household behaviour, it is clear that a family's decisions about fertility fail to take into account the impact of its actions on resources that it does not own. In principle, we could imagine confronting households with a set of prices that fully reflect the social costs of the impact of their actions on the environment and thus cause households to internalize environmental externalities in their economic and demographic decision-making. It is interesting to ask how household fertility decisions would change as/result of facing the 'correct' prices. It is not clear to me that this question has any general answer, because household impacts on the environment are likely to vary with local conditions. One general thing that can be said is a further implication of Engel's Law: namely, that an increase in fertility, holding total resources constant, will tend to increase the fraction of resources devoted to agriculture. For the sake of argument, suppose that we are considering a low-income society in which industrial output is the main alternative to agricultural output, and let us assume that industrial output generates relatively more environmental damage than agricultural output. In this case, confronting households with prices that embody charges for environmental damage will tend to stimulate fertility because it lowers the relative price of children by lowering the relative price of goods most intensively used by children.

In my view, the question of whether higher or lower fertility would be associated with an allocation of resources that appropriately values the local and global environment is a secondary issue. The primary issue concerns obtaining empirically valid measures of environmental values and developing policies and institutions that are sharply focused on mitigating environmental damage. It seems to me that a policy directed simply towards fertility reduction alone is a very blunt instrument in this regard, which could even work in the wrong direction, as is shown in the example discussed in the preceding paragraph.

Economic growth and development undoubtedly create a wide variety of external effects which threaten many aspects of the environment in many different ways. By raising incomes, they also increase the value that individuals place on the services of the environment and, probably, enhance their ability and capacity to create political, social, and economic institutions to deal with such environmental questions. It appears to me to be an empirical question whether, on balance, economic growth and demographic transition tend to increase or decrease environmental problems.

REFERENCES

Arthur, W. B., and McNicoll, G. (1978), 'Samuelson, Population and Intergenerational Transfers', *International Economic Review*, 19: 241–6.

Barro, R. (1974), 'Are Government Bonds Net Wealth?', *Journal of Political Economy*, 82: 1095–117.

Becker, G. S. (1960), 'An Economic Analysis of Fertility', in *Demographic and Economic Change in Developed Countries* (Princeton, NJ), 209–31.

Becker, G. S. (1974), 'A Theory of Social Interactions', *Journal of Political*

Economy, 82: 1063–93.

Becker, G. S. (1981), *A Treatise on the Family* (Cambridge, Mass.).

Becker, G. S., and Lewis, H. G. (1973), 'Interaction between Quality and Quantity of Children', *Journal of Political Economy*, Supplement.

Becker, G. S., and Tomes, N. (1976), 'Child Endowments and the Quantity and Quality of Children', *Journal of Political Economy*, 84: S142–S163.

Bernheim, B. D., Scheifer, A., and Summers, L. H. (1985), 'The Strategic Bequest Motive', *Journal of Political Economy*, 95: 1045–76.

Blandy, R. (1974), 'The Welfare Analysis of Fertility Decline', *Economic Journal*, 84: 109–29.

Cain, M. (1977), 'The Economic Activities of Children in a Village in Bangladesh', *Population and Development Review*, 3: 201–27.

Caldwell, J. C. (1976), 'Toward a Restatement of Demographic Transition Theory', *Population and Development Review*, 2: 321–66.

Cigno, A. (1991), *Economics of the Family* (Oxford).

Coale, A. J., and Hoover, E. M. (1958), *Population Growth and Economic Development in Low Income Countries* (Princeton, NJ).

Cox, D. (1987), 'Motives for Private Income Transfers', *Journal of Political Economy*, 95: 508–46.

Deardorff, A. V. (1976), 'The Growth Rate of Population: A Comment', *International Economic Review*, 17: 510–15.

Deolalikar, A. B. (1993), 'Gender Differences in the Returns to Schooling and Schooling Enrolment Rates in Indonesia', *Journal of Human Resources*, 28: 899–932.

Freeman, R. B. (1986), 'The Demand for Education', in O. Ashenfelte and R. Layard, *Handbook of Labor Economics*, i (Amsterdam), 357–86.

Jones, G. (1990), 'Fertility Transitions among Malay Populations of Southeast Asia: Puzzles of Interpretation', *Population and Development Review*, 3: 507–37.

Kim, O., and Willis, R. J. (1982), 'The Growth of Population in Overlapping Generations Models', mimeo., September.

Knodel, J., Havanon, N., and Pramualratana, A. (1984), 'Fertility Transition in Thailand: A Qualitative Analysis', *Population and Development Review*, 10: 297–328.

Kuznets, S. (1966), *Modern Economic Growth* (New Haven, Conn.).

Lee, Ronald D. (1973), 'Population Size and Real Wages in Preindustrial England: An Economic Analysis', *Quarterly Journal of Economics*, 87.

Lee, R. D. (1980), 'Age Structure, Intergenerational Transfers, Consumption and Economic Growth', *Revue economique*, 31: 1129–56.

Lee, R. D. (1987), 'Population Dynamics of Humans and Other Animals', *Demography*, 24: 443–67.

Lee, R. D., and Lapkoff, S. (1988), 'Intergenerational Flows of Time and Goods: Consequences of Slowing Population Growth', *Journal of Political Economy*, 96: 618–51.

Lee, Y.-J., Parish, W. L., and Willis R. J. (1994), 'Sons, Daughters, and Intergenerational Support in Taiwan', *American Journal of Sociology*, 99: 1010–41.

Leibenstein, H. (1957), *Economic Backwardness and Economic Growth* (New York).

Lillard, L. A., and Willis, R. J. (forthcoming), 'Intergenerational Educational Mobility: Effects of Family and State in Malaysia', *Journal of Human Resources*.

Linder, S. (1971), *The Harried Leisure Class* (New York).

Manser, M., and Brown, M. (1980), 'Marriage and Household Decision-Making: A Bargaining Analysis', *International Economic Review*, 21: 31–44.

McElroy, M., and Horney, M. J. (1981), 'Nash-Bargained Household Decisions: Toward a Generalization of the Theory of Demand', *International Economic Review* , 22: 333–49.

McGarry, K., and Schoeni, R. F. (1993), 'The Measurement of Intergenerational Transfers: A Comparison across Surveys', Health and Retirement Survey, Early Results Workshop, Institute for Survey Research, Ann Arbor, Michigan, September.

Meltzer, D. (1992), 'Mortality Decline, the Demographic Transition and Economic Growth', unpublished Ph.D. dissertation, Department of Economics, University of Chicago.

Menchik, P. (1980), 'Primogeniture, Equal Sharing, and the US. Distribution of Wealth', *Quarterly Journal of Economics*, 94: 299–316.

Mueller, E. (1976), 'The Economic Value of Children in a Peasant Society', in R. G. Ridker (ed.), *Population and Development: The Search for Selective Interventions* (Baltimore), 98–153.

Parish, W. L., and Willis, R. J. (1993), 'Daughters, Education and Family Budgets: Taiwan Experiences', *Journal of Human Resources*, 28: 863–98.

Samuelson, P. A. (1958), 'An Exact Consumption Loan Model of Interest With or Without the Social Contrivance of Money', *Journal of Political Economy*, 66: 923–33.

Samuelson, P. A. (1975), 'The Optimum Growth Rate for Population', *International Economic Review*, 16: 531–8.

Schultz, T. P. (1990), 'Testing the Neoclassical Model of Family Labor Supply and Fertility', *Journal of Human Resources*, 25: 599–634.

Schultz, T. P. (1992), 'Demand for Children in Low Income Countries', mimeo., prepared for M. R. Rosenzweig and O. Stark (eds.), *Handbook of Population Economics*.

Solow, R. M. (1956), 'A Contribution to the Theory of Economic Growth',

Quarterly Journal of Economics, 70: 531–38.

Strauss, J., and Thomas, D. (1993), 'Human Resources: Empirical Modeling of Household and Family Decisions', mimeo., prepared for T. N. Srinivasan and J. R. Behrman (eds.), *Handbook of Development Economics*, iii.

Thomas, D. (1990), 'Intra-Household Resource Allocation: An Inferential Approach', *Journal of Human Resources*, 24: 635–64.

Thomas, D. (1993), 'The Distribution of Income and Expenditure within the Household', *Annales d'économie et de statistiques*, 29: 109–36.

Weiss, Y., and Willis, R. J. (1985), 'Children as Collective Goods and Divorce Settlements', *Journal of Labor Economics*, 3: 268–92.

Wildasin, D. E. (1990), 'Non-Neutrality of Debt with Endogenous Fertility', *Oxford Economic Papers*, 42: 414–28.

Willis, R. J. (1973), 'A New Approach to the Economic Theory of Fertility Behavior', *Journal of Political Economy*, Supplement: S14–S64.

Willis, R. J. (1980), 'The Old Age Security Hypothesis and Population Growth', in T. Burch (ed.), *Demographic Behavior: Interdisciplinary Perspectives on Decisionmaking* (Boulder, Colo.).

Willis, R. J. (1982), 'The Direction of Intergenerational Transfers and Demographic Transition', *Population and Development Review*, Supplement, 8: 207–34.

Willis, R. J. (1987), 'Externalities and Population', in D. Gale Johnson and R. D. Lee (eds.), *Population Growth and Economic Development: Issues and Evidence* (Madison, Wis.), 661–702.

Willis, R. J. (1988), 'Life Cycles, Institutions, and Population Growth: A Theory of the Equilibrium Interest Rate in an Overlapping Generations Model', in R. Lee, W. B. Arthur, and G. Rodgers (eds.), *Economics of Changing Age Distributions in Developed Countries* (Oxford), 106–38.

Willis, R.J. (1993), 'A Theory of Out-of-Wedlock Childbearing', mimeo, presented at Workshop on Low Income Labor Markets, Institute for Research on Poverty, University of Wisconsin, 22–6 June 1993.

World Bank (1990), *World Development Report* (Oxford).

World Bank (1992), *World Development Report* (New York).

NOTES

1. See McNicoll (Chapter 8 of this volume) for further description in patterns of demographic and economic transitions across broad regions of the world.

2. In this chapter, I do not attempt to deal with 'intrinsic' values of natural and biological resources beyond their value to people.

3. It would be more accurate, however, to say that the factors embodied in the

Malthusian model have been largely irrelevant in determining the course of population growth since the Industrial Revolution because their influence has been swamped by other factors associated with the cumulative advance of knowledge that were beyond Malthus's imagination. The Malthusian model has been shown to work impressively well by Lee (1973) over a 500-year span in pre-industrial England. More recently, Lee (1987) has shown that density-dependent homeostatic feedback, which is the most distinguishing feature of the Malthusian model, operates only weakly over long periods of time.

4. The use of immigration rather than fertility in the example enables us to avoid inter-temporal questions which would add complexity without changing the analysis. See Willis (1987: 672–4) for a more detailed analysis of Blandy's argument.

5. The fact that children do not choose to be born raises some interesting and difficult questions in welfare economics, but these primarily concern the extent to which the society or the polity believes that parents can be entrusted to look after the well-being of their children rather than issues about externalities. Many societies impose some legal and normative constraints on reproductive behaviour (e.g. legislation about abortion, sanctions on out-of-wedlock child-bearing) and on their upbringing (e.g. compulsory schooling, child-labour laws). Within such constraints, however—and with notable exceptions such as the Chinese 'one-child policy'—parental decisions tend to be regarded as sovereign. Thus, I believe that most people would hesitate to say that the child of another person would be better off not being born rather than being born into a very poverty-stricken household. For example, Carolyn Bledsoe's discussion of African fertility in Chapter 5 above is quite explicit on this point. In contrast, most people are quite comfortable with the notion that parents may ethically choose not to have a child because they do not wish to deny themselves or the potential child's siblings some luxury goods such as a fancy house or regular skiing holidays.

6. Samuelson went on to define 'the goldenest of golden rules' by finding the jointly optimal rates of capital accumulation and population growth that would maximize the level of per-capita economic welfare in a steady-state growth path. To the delight of subsequent generations of graduate students in economics, he actually found the conditions that would minimize per-capita welfare (Deardorff 1976). Subsequently, Kim and Willis (1982) provided necessary conditions for the existence of an optimal rate of population growth in Samuelson's sense. This is not a useful concept of optimal population growth, however, because it omits any consideration of the utility that parents receive from bearing children.

7. Assuming that the length of a generation is twenty-five years, this cor-

responds to an annual rate of population growth of 2.8%, which is about equal
to average rate of population growth during the 1980s in those countries clas-
sified by the World Bank as at the low or low-medium level of development.

8. A golden-rule equilibrium is one which selects that savings rate which max-
imizes per-capita utility in steady state.

9. The average age of consuming is given by

$$a_c = \int xs(x)c(x)e^{-gx}dx \,/\, \int s(x)c(x)e^{-gx}dx$$

where x is age, $s(x)$ is the probability of survival to age x, $c(x)$ is age-specific
consumption, and g is the rate of population growth. Note that the above
expression would yield the average age of the population if $c(x)$ were constant across all ages. The average age of consuming, therefore, gives a con-
sumption-weighted average age of the population. The average age of
producing is defined analogously, replacing age-specific consumption by
age-specific labour productivity in the above expression.

10. There is evidence of such a relationship for the United States. Specifically,
Freeman (1986: table 6.2) presents data showing a strong positive correlation
between the educational attainment of workers in different sectors of the economy and employment growth in the sector between 1960 and 1970. For
example, employment in the most highly education-intensive sectors such as
professional services or finance, insurance, and real estate grew by about 40%
over the decade, while the least education-intensive services such as agriculture and personal services actually fell. Across the twelve sectors reported by
Freeman, the rank correlation between education-intensity and employment
growth is about 0.8 by my calculation.

11. Economic development tends to generate increased investments in human
capital for other reasons as well. As one example, mortality decline increases
the incentive to invest in human capital (Meltzer 1992). As another, the rising
fraction of the population residing in urban areas makes it easier to provide
the bulk of the population with easy access to schools.

12. Additional considerations are needed to account for the timing of monetary
transfers from parents to children. For example, end-of-life bequests provide
a means by which parents and children may share the risks associated with an
uncertain length of life; they also add to the parents' bargaining strength by
giving them the 'final move' in bargaining with their children (Bernheim,
Scheifer, and Summers1985). Conversely, if children face borrowing constraints, *inter-vivos* transfers to help finance the purchase of a home or to start
a new business may be preferred to bequests.

13. Even when parents desire to make monetary transfers to their children, there
may be conflicts of interest concerning care and attention that elderly parents
wish to receive from their grown children and the amount that children wish
to provide. See Bernheim, Scheifer, and Summers (1985) and Cox (1987) for

models of inter-generational relationships which emphasize such conflicts.

14. Analyses of probate records in the United States generally fail to support this prediction of the Becker–Tomes model. Most typically, estates tend to be divided equally among the children, even when the children have quite unequal incomes (Menchik 1980). Preliminary analysis of some new data on *inter-vivos* transfers does appear to show some 'compensatory' patterns (see McGarry and Schoeni 1993).

15. See e.g. Mueller (1976). However, using data from Bangladesh, Cain (1977) finds that male children become net producers between the ages of 10 and 13 and that, by age 15, their cumulative production will have exceeded their cumulative consumption. This implies that male children provide a positive expected rate of return, but it is not clear whether it exceeds expected rates of return on alternative investments. In any case, as Cain stresses, the dominant economic benefits from children occur because of the insurance against risks that they provide.

16. Using a rich body of survey data from Malaysia, Lillard and Willis (forthcoming) attempted to discover whether the NEP programme had influenced the educational attainment of children, after controlling for a large number of variables measuring family background, resources, location, and the availability of schools. They concluded that there was a discernible effect on racial differences in educational attainment, but that the magnitude of the effect was not large.

7

Government, Population, and Poverty: A Win-Win Tale

Nancy Birdsall[*]

Traditional concern that rapid population growth slows development in poor countries has been reinforced in recent years by a growing awareness of the fragility of the natural environment in poor countries and its apparent vulnerability to the stresses associated with increasing population size, whether in a growing or a stagnating economy. But reinforced concern has not led to any consensus about whether governments should intervene aggressively to reduce population growth rates. At least some careful analysts remain sceptical that rapid population growth in and of itself is a fundamental cause of either slower economic growth or of natural resource problems in developing countries.[1] Others, even if confident that rapid population growth is a problem, worry that government intervention will threaten the sanctity of individual and family decision-making in the sensitive arena of reproduction, and question whether the future and uncertain benefits of a smaller population would outweigh immediate welfare costs—in terms of human pain and suffering—of strong policies to reduce fertility. Such fundamental questioning is behind the international opprobrium China has suffered because its birth-control policies have brought such human pain.

In this chapter I first summarize and discuss the three principal concerns raised by rapid population growth in developing countries: slower economic development, greater environmental damage, greater poverty and income inequality. I then link these concerns to specific rationales for

* Inter-American Development Bank. This chapter draws in part from Birdsall and Griffin (1993) and Chomitz and Birdsall (1991). I am grateful to Robert Cassen for good ideas and general encouragement.

government intervention to reduce rates of fertility, and show the basis for these rationales in simple welfare theory (that is, economic theory which starts from the notion that the objective of economic arrangements is to maximize human happiness). Finally, I discuss the kinds of public-policy interventions that these rationales justify. (At the outset it is worth noting that not all 'interventions' need imply more government-funded programmes or more government involvement in the economy; some might consist of government permitting or encouraging market-led outcomes, for example, reducing regulations that stifle private-sector supply of contraception.[2])

In proceeding from concerns, to rationales, to interventions, I do not systematically review the empirical evidence linking rapid population growth to the three concerns (though I do allude to some such evidence). My interest is not in the merits of that evidence but in whether such evidence, were it correct, would justify government intervention (governments, after all, may do better to leave problems alone rather than intervene and make matters worse); and in what kinds of interventions would be justified. As should become clear, this approach (not assessing the merits of the evidence) is more sensible than it might seem—because the kinds of interventions that are justified turn out to be 'win-win', that is, they can be justified independently of the population question.

The key reason for this convenient result is the link between high fertility and poverty. Poverty is both a root cause and a common outcome of high fertility. A massive literature amply documents the many specific characteristics of poor households that contribute to high fertility— high infant mortality, lack of education for women, too little family income to 'invest' in children, leading to parents having many children rather than concentrating investments in a few, and finally, for many poor couples, poor access to contraception of reasonable cost and quality.[3] The generalized effect of poverty on fertility is demonstrated in almost every developing country in the form of large differentials between the average family size of households, depending on income, education, and other variables that measure or reflect poverty. At the same time, poverty is often an outcome of high fertility; as discussed below, large family size strains family budgets and reduces the ability of families to invest in their children's health and education.

There is, in short, a vicious circle of high fertility and poverty in many developing countries. So reducing fertility levels in poor countries requires change in the conditions of poverty which cause high fertility in the first place. Thus, we are faced with the need to define interventions which will not only reduce the fertility of the poor, but do so by making the poor better off. Such interventions need not pass so strict a test of justification and acceptability as 'population' interventions which involve a possible trade-off between the welfare of the poor and less-rapid population growth, or between the welfare of the poor now and the welfare of future generations. This convenient result thus inspires the title of this essay: Government, Population, and Poverty: A Win-Win Tale.

A second consideration arises as background. Institutional constraints to the ideal policy regime cannot be ignored. In some cases, institutions (including political, administrative, and social) make it difficult to initiate and implement non-population policies which would mitigate or even eliminate the worst effects of population growth. An example is better pricing of such common property resources as forests—which would greatly reduce logging and burning independent of rapid population growth. The institutional constraints to ideal policies make recourse to fertility reduction policies, especially if they are win-win, easier to justify.

1. THREE PROMINENT CONCERNS ABOUT RAPID POPULATION GROWTH

Three concerns about rapid population growth in developing countries are:

- that rapid population growth reduces the rate of economic growth by reducing investments in human capital, investments that have powerful effects on economic growth because of positive externalities;
- that rapid population growth itself has negative externalities for the environment, leading in some scenarios to degradation of natural resources at the local and national level and contributing to such global problems as loss of biodiversity and the possibility of global warming;
- that rapid population growth has negative 'pecuniary' externalities, that is, that it reduces the incomes of some groups, particularly the

poor, compared to other groups, and therefore exacerbates the problems of poverty and income inequality in developing countries.[4]

A key word in the statement of each concern is 'externality'. An externality exists whenever the costs and benefits of an activity are not fully internalized by the actor. If my activity imposes costs on my neighbour which I do not bear (I emit smoke or noise and pay no fees), or provides benefits I cannot fully capture (immunizing my child prevents another child's sickness), there exists an externality. Where a negative externality exists, if every individual does what he or she wants (many polluters), the results for all as a group may be unsatisfactory. The good of each does not harmonize with the good of all, and each would be better off if others did not all act as she or he acts. The externality signals a failure in the way markets are working, and justifies some collective or government intervention—in the case of pollution, a tax on emissions to impose on each polluter the full social costs of his pollution, and in the case of immunizations, a subsidy to encourage more parents to obtain immunizations.

Concern about population growth is often based, implicitly or explicitly, on the idea that there are negative externalities to child-bearing in one or more of the three forms set out above: my decision to have a child imposes costs on society that I do not take into account.[5] Despite the popularity of this assumption, there has been relatively little theoretical work[6] and almost no empirical work describing the nature and magnitude of externalities.[7]

Conceptually, the potential for externalities to arise is clear. First, when there is public national wealth (for example, state-owned lands or mineral deposits), the birth of a new citizen dilutes everyone else's claim on this jointly owned property. Similarly, if there are non-privatizable, congestible common property resources, such as fisheries and aquifers, an additional birth reduces per-capita consumption of the resource. If the resource is not properly managed, as in the classic 'tragedy of the commons' scenario, in which access to the resource is unlimited, then population growth exacerbates the losses due to mismanagement. The latter problem is at the heart of the concern about the effects of population growth on the environment.

A formally similar situation occurs when citizens are entitled to public-transfer payments (social security) or subsidized services (education, health), consume public goods, and are obliged to pay taxes. In such situations, the birth of an additional citizen adds to both public revenue and expenditure, resulting in positive and negative externalities to other taxpayers. Depending on the amounts and life-cycle timing of tax payments and receipt of benefits, net externalities could be positive or negative. In societies with high rates of fertility, the assumption is that the externalities are negative—that the social costs of educating more and more children will not be recouped eventually in higher taxes.

A pecuniary externality to child-bearing can occur in the labour market. As population, especially rural population, increases, the wages of labour go down, and rents increase. For individual landless or land-poor labourers it is rational to have many children as a strategy for maximizing old-age income. When all pursue this strategy, however, wages are depressed, and both the parents' living standards and those of their children are reduced below their expectations. Hence, one poor family's decision to have children imposes costs on its peers; if all joined in a labour-supply cartel and agreed to have fewer children, the group's income would increase.

This is not a 'true' externality because it operates through the market. Although wages are depressed by population growth, rents are boosted; high fertility is potentially Pareto-superior to low fertility, assuming that landlords redistribute some of their rents. In practice, however, the landlords or capitalists are unlikely to redistribute their rents. So where income inequality and absolute poverty are already high, population growth is likely to compound these problems.[8]

A final source of child-bearing externalities occurs when couples care about their relative—rather than absolute—family size. A family's voice in the village council, prestige, physical security, or claim to the use of common property may depend on its size relative to the group (Crook 1978). When a couple has more children, they slightly increase the average family size in their group, making others relatively worse off. An attempt by each couple to have more children than the average family size is mutually frustrating: average family size increases, and no couple achieves the sought-after advantage, although all now have larger families to support.

Consider the growth, environment, and income distribution concerns in terms of these child-bearing related externalities.

Negative Growth Effects

The typical growth concern is that rapid population growth reduces economic growth by reducing investment in physical and human capital.[9] This broad growth concern has been amply explored for at least two decades; its merits are difficult to address empirically. In the absence of compelling evidence it remains a matter of controversy.[10] On the one hand, economists have noted that, in the absence of externalities, a negative effect of rapid population growth on economic growth need not in itself be of great concern. Parents may fully realize that children are costly to them (and to society), and yet prefer to have more children rather than higher consumption of other costly things.[11] If so, rapid population growth may be socially optimal even if it impedes economic growth.

However, a countervailing argument can be made. Recent approaches to growth theory emphasize the possibility of important positive externalities for economic growth associated with investments in human resources.[12] In cross-country empirical studies of growth, measures of the educational attainment of populations have been consistent and important explanators of growth success; this in itself is not new. More important, recent studies indicate that education matters over and above its effect as an additional input to production; at the country and firm level, it is also associated with higher total factor productivity, that is, with higher product for given inputs.[13] Any persistent effect of education on economy-wide productivity would suggest a positive social externality of a more educated work-force; then, to the extent that rapid population growth inhibits public and family investments in education, its effects on growth would be negative—and furthermore would not be socially optimal.

At the family level, there is evidence that high fertility inhibits investments in children's education. A study of families with twins in India found that the additional unexpected child represented by twins reduced enrolment levels of all children in the household.[14] Estimates based on Malaysian data show that couples with a higher biological propensity to have births (higher fecundity or supply of births) are also characterized by lower schooling attainment for their children (Rosenzweig and Schultz 1985).

At the economy-wide level, there is no simple evidence that rapid population growth reduces overall effort to invest in children, measured, for example, by proportion of GDP spent on education. That proportion does not vary substantially across countries (falling between 2 and 4 per cent in almost all developing countries). However, it is obvious that the same effort so defined has different results depending on the proportion of children in a population; only if a society succeeds in increasing its effort or increasing its efficiency in use of given resources as the size of its school age cohorts grow, can it keep up. Evidence of declining quality of education in Africa and Latin America suggests increasing difficulties in doing so. For example, the absolute size of the school-age population in Mexico increased by 60 per cent from 1970 to 1989; in Korea it remained virtually stable. Expenditures in Mexico increased as well—but by 60 per cent, just enough to maintain per child spending.[15]

In short, high fertility does seem to be associated with less education, and at the economy-wide level less education has a social cost—lower economic growth. Rapid population growth thus implies the loss of a potential positive externality of education for economic growth.[16] Put another way, parents who have many children, and educate them less, in effect deprive society of the potential extra-productivity effect of a more-educated population.

Negative Environmental Externalities

The environmental concern about rapid population growth is straightforward. For given levels of consumption, more people imply more stress on natural resources, including both sources (forests, water) and sinks (the air which receives pollution). In the absence of prices which reflect the true scarcity value of these sources and sinks (and recognizing the institutional and political difficulties of, for example, adequately pricing for water, for clean air, and for use of common natural resources such as many forests and oceans), there is likely to be excessive consumption of these 'goods' from society's point of view. Excessive consumption is multiplied the more people there are.

These potential negative environmental externalities, at both the local and global levels, are increasingly invoked as a rationale for efforts to reduce population growth in developing countries.[17] At the local level, the combination of poverty and rapid population growth is often cited as contributing to environmental degradation—for example, because population pressure leads to farming of hillsides and other marginal areas (causing soil erosion) or heavy cutting of forests for fuel (causing damage to watersheds, hence to agriculture, and contributing to possibly irreversible reductions in biodiversity).[18]

At the global level, though the fossil-fuel emissions that currently contribute to possible global warming are much greater in industrialized countries on a per-capita basis, future rates of income and population growth are likely to be greater in developing countries, implying that the contribution of these countries to global emissions could rise from about 20 per cent today to 50 per cent by the middle of the next century (Birdsall and Dixon 1991; Bongaarts 1992).

Concern about the effects of rapid population growth on global food supplies arises now not because of absolute food shortages in the near term, but because necessary increases in either the intensive or the extensive margin of agriculture are likely to impose unsustainable stress on the natural resource base. In relatively poor settings in many developing countries, increased demand for food (and for farm income) with rising population has contributed to the extension of farming to less-productive and usually more environmentally fragile areas—as in the case of clearing of forests and of hillsides,[19] or to what is viewed as unsustainable recourse to chemical fertilizers. Some analysts now believe that the problem is not just a local one—that in the future sustaining food availability at acceptable environmental and economic cost will also be an issue at the global level. One estimate is that between now and the year 2030 demand for cereals will nearly double, from 2 to 3.6 billion tons, and that 90 per cent of this projected demand will be due to rising population, and only 10 per cent to rising incomes (Cassen 1993). In a thorough analysis of prospects, Crosson and Anderson (1991) conclude that this demand cannot be met on the basis of existing knowledge and available technologies without considerable environmental damage. New knowledge, as well as greater use of existing technologies and best practice, is therefore critical. In developing areas, the critical constraint is likely to be water, for which there is increasing competition from non-agricultural sources.

Worsening of Income Distribution and Poverty

Thirdly, there is the distribution concern—that is, the concern that rapid population growth creates the 'pecuniary' externalities described above in the labour market; and the related but important concern that their own high fertility hurts the poor.

Even if rapid population growth does not create pure negative externalities and thus social losses at the aggregate level, it may systematically lead to lower levels of income of the poor compared to the rich. At the aggregate level, rapid population growth increases the availability of labour in an economy relative to land and physical capital, reducing wages.[20] This is likely to worsen inequality and hurt the poor, who are more reliant on labour income. In most economies today, the distribution of income is in fact more closely linked to differential returns to different levels of skills among labour than to differences between labour and capital. For this reason, the effect of different rates of population growth on differences in income by education level provides insight on the effects of rapid population growth on aggregate income distributions. From this point of view, evidence from Brazil that unskilled, but not skilled, labourers suffer a relative decline in wages if they are members of a large cohort is indicative. The size of cohort (and implicitly the rate of population growth at the time a worker was born) matters more (in a negative way) for the less educated than for the more educated.[21]

The probable deleterious effect of rapid population growth on the distribution of income has not been given much attention. However, there seems little doubt that the sustainability of the economic and political reforms of the last decade is in part a function of the extent to which economies and polities succeed in ensuring that the costs of reform are not unduly borne by the poor, and that the benefits of growth, when it returns, are widely shared.[22] The experience of the high-growth countries of East Asia demonstrates this point amply; there, growth was widely shared, and the brief periods of adjustment and austerity were also shared.[23] Other cross-country studies also suggest that the distribution of income affects growth, presumably via its effect on the capacity of governments to implement and sustain economic policies that are costly to at least some groups in the short run. To the extent that rapid population growth exacerbates income inequality, it may thus contribute to the difficulty of sustaining growth, and in particular to the difficulty of sustaining a process of economic reform to ensure growth.

The effect of high fertility (the underlying cause of rapid population growth) on poverty, independent of effects on aggregate income distribution, is easier to document. High fertility contributes to poverty at the family level by straining the budgets of poor families and by reducing available resources to feed, educate, and provide health care for children. The evidence cited above regarding the effects of twins on family spending per child for education speaks to this effect.[24]

High fertility in poor families need not reflect irrational decisions on the part of poor parents, even though it reduces family resources per-capita in the short run. On the contrary, it can reflect reasonable decisions on their part—to ensure greater future family income once children start working, or to ensure their own security in old age via support from their children. As noted above, it may also reflect parents' decisions to enjoy additional children rather than other forms of consumption. In both cases, however, children themselves are likely to receive less health and education—implying a negative inter-generational externality, that is, a situation in which the parents do not incur the full costs of high fertility but pass some of those costs on to their children. In some cases, parents also may end up worse off—if more children than hoped are girls (who will provide less support in old age), or if children turn out to be less helpful, less capable, or less pleasant than anticipated. This only implies that some parents lose what are reasonable gambles. It is, of course, also possible that poor families have more children than they want, because they have poor access to modern contraception; or they have more children than they might have chosen, because they had poor access to information about the increasing probability their children would survive or the increasing opportunities for and returns to education. I return to these issues of inter-generational welfare, choice, and information in Section 3 below.

2. RATIONALES FOR GOVERNMENT INTERVENTION

Do these concerns justify government efforts to reduce fertility?

First, should governments encourage lower fertility as a way to capture the positive externalities for economic growth of resulting higher private and public investments in education? A purist would argue: not necessarily. The appropriate response is to subsidize education (and other investments in human capital, including health) directly—enough to raise private consumption (which in this case is also private investment) to the socially optimal level.

But there are a number of practical problems with the purist response. First, virtually all governments subsidize education already. Current subsidies are limited not by some notion of an optimal ceiling that has been reached, but by institutional realities—fiscal constraints and the administrative difficulty of subsidizing any service efficiently (whether subsidies are provided directly or through private providers). Secondly, in some settings higher subsidies to education might actually induce higher fertility by reducing the cost of children to parents without altering the benefits, thus creating another round of need for new subsidies! Finally, even large subsidies to education (and health) may be insufficient to induce the socially optimal private investments in these forms of human capital, since some commitment of children's time, which may be valuable to parents, is required.

In short, it is usually less costly and more efficient for societies to encourage education by combining direct subsidies to education with other separate measures to reduce family size—rather than relying on the former alone.

Should governments encourage lower fertility in order to reduce negative environmental externalities? Again, a purist might argue no: that the appropriate response would be for government to impose higher prices on environmental goods—charging for burning of tree cover, for fishing licenses, and for pollution via chemical fertilizers. Again, however, institutional constraints matter; adequate pricing of environmental goods is politically and administratively difficult, even in rich countries. While pricing should be tried, the existence of the problem also justifies collective efforts to reduce fertility. If a country can, through some policy change or programme, induce parents voluntarily to have fewer children, then environmental degradation can be slowed and food production per person maintained, and the country as a whole made better off. Analogously, in a global society, if industrialized countries can induce developing countries to reduce their fertility levels, then global welfare can increase.

Do negative effects of high fertility on income distribution and on poverty also justify government intervention? In theory the answer depends on whether a particular society cares about the existence of poverty among some of its members, and about the distribution of income. In fact, most societies do seem to care. First, the utility of both the poor and the rich is probably affected by knowledge that the other group is vastly better or worse off. Secondly, as noted above, the political sustainability of economic reforms and thus of economic growth may be jeopardized where income inequality is great, harming everyone. Thirdly, democracy, a political arrangement which probably improves human welfare relative to other arrangements (because such freedoms as those of speech, religion, and association make people better off) may be difficult to sustain where income inequality is high and social immobility low.

Consider a typical developing country in which the poor have higher fertility than the rich. Suppose a social programme—for example, education or family planning—could reduce the number of children born to the poor while making them no worse off than before with their larger families. The social programme is financed by taxing the rich. Through such a programme, both the rich and the poor can enjoy an unambiguous increase in welfare—the rich because they prefer that the poor have smaller families, the poor because they benefit from the social programme.

In summary, the three concerns all justify in principle government intervention, because each concern is rooted in a market failure which collective action through government could correct. Interventions specifically aimed at reducing fertility are among those that can be justified, as long as they enhance rather than reduce the welfare of the poor.

A Note on Wrong Prices as Externalities

Dasgupta (1993) has pointed out a connection between the population problem and the under-pricing of firewood. Firewood in developing countries is under-priced because it is a common property resource, available at no direct cost from local forests. Its under-pricing (it is free except for the opportunity cost of the labour in collecting it) reflects the lack of a market in common property resources. Collecting firewood is a common task for children in poor families; the low price of firewood, and its increasing scarcity (in part because of its low price), make children more valuable to parents than they would be otherwise. This 'wrong price' of firewood in fact reflects a market failure (or externality problem) which contributes to high fertility, and makes high fertility in turn a contributing factor to the worsening of the original environmental problem of under-priced firewood. In this way, rapid population growth results from and compounds a 'wrong price'.

Other wrong prices, which reflect externality problems, can also increase fertility and population growth. In West Africa the full costs of children are not borne by their parents. Other relatives are expected to provide support, including through extensive fostering-out (Bledsoe, Chapter 5 of this volume); because parents can capture many of the benefits of children, but need not bear all the costs, there is a kind of externality. The same problem arises wherever the husband is the principal decision-maker and can capture many of the benefits of children yet pass on many of the costs to his wife. Finally, wherever social norms (what Dasgupta refers to as 'diffuse externalities') or local conditions mean that families with more children than average capture some benefits (as mentioned in Section 1) but can pass on to society some costs, there is a 'wrong price' for children rooted in an externality. All these cases of wrong prices in fact justify in principle some collective action or public policy or intervention designed to reduce fertility.

3. WHAT KIND OF INTERVENTIONS?

The solution to externalities implies the need to ensure that actors internalize fully the social costs of their actions—in the case of fertility, that the poor who have high fertility bear the full social costs of their choice. At the same time, the poor have many children in part because they are poor, so reducing this fertility requires a change in the conditions of poverty itself—it requires that the poor be made better off. In short, we are in search of interventions which meet the following apparently difficult criterion: *interventions which simultaneously increase the cost of children to the poor, inducing lower fertility—and improve the welfare of the poor.*[25]

Consider five types of intervention. Except for the first, each does meet the criterion.

Taxes on Children and Incentives to Reduce Fertility

If child-bearing imposes external costs, the standard economic remedy would be to impose a tax on children equivalent to the external costs. Then, assuming that couples have perfect control over their fertility, a child would be born only if its value to its parents exceeded its total cost to society.

Although taxes in one form or another have been used—in China and Singapore—they are problematic from a welfare point of view because they are likely to hurt the poor. For the poor much more than for the rich, children are likely to be an important source of old-age income and security. Yet in no society—developed or developing—is it easy to borrow against the future earnings of one's children. Because of this failure of capital markets, it is easy to imagine a tax forcing a peasant to forgo the birth of a child whose financial security value alone is much greater than the tax. Even if capital markets posed no problem, an administratively simple tax would probably be highly regressive, and it might be difficult to guarantee that the tax revenues would be progressively distributed. There is a particular danger that a tax would penalize the children born, by accident or design, into large poor families.

A tax-financed incentive for not bearing children would have the same effect as a tax, without apparently hurting the poor—since in principle an incentive would add to income. However, incentives, though also used in China, as well as India and Bangladesh, are politically unpalatable in many countries[26]—in part because they could pose a 'tragic choice' for the poor, inducing decisions which might later be regretted.[27]

Education and Other Social Programmes to Raise the Cost of Children

Social programmes that promote investments in the human capital of the poor are consistently associated with reduced fertility among the poor. Probably the single most important is expanded educational opportunities for girls. These raise the cost of children to women once they are of child-bearing age, by raising the opportunity cost of the potential mother's time, both in the labour market and in other non-labour activities. Because children are costly in time, they are more costly to mothers whose time is more valuable. So educating girls raises the cost of children to future mothers, reducing fertility while at the same time improving the income and other prospects of those girls.

Education of girls as well as boys can also have an immediate effect on current and potential parents. Where schools are inaccessible or of such low quality that children learn little, they do not affect the 'cost' of children to parents. However, the appearance of educational opportunity in a village or neighbourhood, in the form of an accessible school of reasonable quality, can raise the cost of children by reducing children's potential time on household and family chores.[28] By reducing the cost of education and thus the cost of producing an educated child, schools can actually raise the cost of having many children. Adding a school to a village is obviously also consistent with improving the welfare of the poor in that village.

Once a school exists in a village, raising tuition fees or other direct costs (for uniforms, books, equipment) would further raise the cost of children to parents—and is clearly consistent with the notion of ensuring that parents bear more of the social costs of their children. Without other changes, however, such as compensatory changes in the quality of the school, or the redistribution of the resulting resources in the form of transfers back to the poor, raising fees could reduce the poor's welfare— either by reducing consumption or by the perverse effect of inducing withdrawal of children from school.

Pension and other programmes that provide some form of security in old age, or that otherwise improve access of parents to capital markets, reduce the benefits of children as providers of old-age security, and, all other factors being equal, raise their relative costs. Like many social programmes, these are also generally associated with increases in the welfare of the poor.

All these changes in the parents' situation affect demand for children by raising the 'price' of children to parents.[29] These programmes tend to have the same effect as taxes on children, but they work indirectly and have the desirable characteristic of succeeding by raising the 'quality' of children.

Adjusting the Price of Contraception

If high fertility has negative externalities for societies and economies as a whole, then it makes sense to ensure that unwanted fertility (of the rich and poor alike) is minimized.

Is there unwanted fertility in developing countries? In many countries significant proportions of women say that they would like to limit their fertility, yet do not do so. This condition of 'unmet need' characterizes a quarter of married fecund women in Egypt, one-fifth in Tunisia, one-eighth in Brazil (Westoff and Ochoa 1991). A good reason for 'unmet need' may be the high price, in one form or another, of contraception. (Demographers use the term 'unmet need' to signify would-be demand for contraception were the price of contraception zero, including psychic as well as monetary costs.) Chomitz and Birdsall (1991) set out a number of market failures that raise the 'price' of contraception and may explain why individuals fail to use contraception even when they want no more children:

(*a*) *Failure in the market for information.* If information (for example, about the possibility of controlling fertility) cannot easily be tied to a specific marketable product, the market may not provide it. This is one reason governments often finance agricultural extension services. Similarly, information about rhythm and withdrawal has no private market; nor does information about the contraceptive pill in rural areas where there is no real market for private medical services. Moreover, consumers may not seek new but costly information, especially on contraception, since the product is a complex set of ideas and procedures whose perceived risks are high.

(*b*) *High costs of early adoption of contraception.* It is easy to imagine a situation in which each family would find fertility limitation advantageous, were it not for the fear of violating social norms and incurring social sanctions. In such a situation, publicly provided information or subsidies could help change norms.[30]

Similarly, early adopters of contraception lower the costs of subsequent adoption for others by providing reliable and specific information about the risks and benefits of contraception. To the extent that family planning is regarded as particularly risky, the demonstration effect, once pioneers verify the innovation's efficacy and safety, could be particularly important.

(c) *Failure of credit and insurance markets.* If the benefits to a couple of a smaller family—reduced expenditure on food, more time to devote to work or leisure, increased maternal attention per child, reduced health risks—substantially outweigh the costs of adoption of contraception (and the pleasures associated with a larger but poorer family), why does not the couple itself finance the costs of contraception? There are two possible reasons. First, many poor couples, lacking collateral, are not creditworthy. Secondly, couples may be able to finance the average cost of adopting contraception, but could not afford, and in the absence of insurance markets, cannot insure against, the possibility of medical complications entailing a long period of unemployment.[31]

These credit, information, and other constraints that raise the 'price' of contraception tend to affect the poor more than the rich, and of course the direct costs of contraception for the poor represent a larger fraction of the total resources, both human and financial, available to them. In the face of these costs, the most direct way for government to encourage use of contraception is to subsidize family-planning services for the poor, as in fact many countries do. Subsidized family-planning services targeted to the poor can, assuming any unwanted fertility or 'unmet' demand, both reduce fertility and, by increasing choices for poor couples (and especially for women), improve the poor's welfare.[32]

Policy or government failures may also explain some of the unmet need for contraception. All countries regulate access to modern contraception. Some forms of chemical and hormonal contraceptives and of abortifacients are banned; many others are available only through a medical prescription (despite good evidence that trained paramedics can, for example, screen effectively for medical contra-indications to use of the pill). Advertising is often restricted. Though some of these restrictions represent efforts to ensure safe use, others arise from long traditions of medical control or from incorporation into government regulations of religious or cultural views that not all members of a population necessarily share. In short, there are settings where governments need not positively subsidize contraception, but only reduce current restrictions to its use through appropriate deregulation. The widespread use of modern contraception by the poor in many countries of Latin America provides a good example of how, despite the constraints noted above, there can be a successful market supply without large public subsidies.

Improving Information

Information is costly—and more costly to the less educated. Poor couples in developing countries may be unaware of changes in factors that are critical to their fertility decisions. Higher returns to education in a rapidly growing economy might mean that having fewer children in order to educate them better makes sense. Recent declines in child mortality might mean that fewer pregnancies are needed to ensure a certain family size. Greater mobility of children as urbanization proceeds might mean that parents might not benefit from additional children as security in old age. To the extent that any of these factors apply, government policies to educate and inform the poor about the changing world might reduce fertility, and would, in any event, make the poor better off.

Economic Management for Broad-Based Growth

None of the programmes and policies set out above should obscure the central relevance of good public management of the economy in reducing fertility. Good management includes implementing policies and programmes which ensure broad-based growth, that is, growth which benefits all groups, including the poor. As noted repeatedly above, poverty is a root cause of high fertility. Growth which raises incomes of the poor will lower fertility.

There are many ways in which poor economic policies penalize the poor and encourage high fertility: poorly functioning capital markets, inflation, and other sources of household economic insecurity make children one of the few viable assets a poor couple can count on; overvalued exchange rates reduce the local-currency value of farmers' exports, import substitution policies penalize labour, restrictions on access to foreign exchange penalize small entrepreneurs, public monopolies of industries and banking mean higher prices for poor consumers, and so on. Broad-based growth has reduced poverty dramatically in such economies as Indonesia, Thailand, Botswana, Chile, and Costa Rica. The keys to broad-based growth, based on the success of East Asian economies, include universal access to basic education (which not only reduces fertility but enhances productivity in the work-place and at home); avoidance of undue taxes on the agricultural sector, including overvalued exchange rates and other trade restrictions which burden agricultural exports; and emphasis on non-traditional exports, which encourage productive adaptation and innovation and tend to be labour-demanding.[33]

Outside East Asia, only a few developing countries have consistently adhered to policies of broad-based growth; many have instead, implicitly or explicitly, managed the economy for the benefit of various interest and élite groups: urban consumers, organized workers in the formal sector, civil servants, users of subsidies to capital, and so forth. Broad-based growth is not necessary to fertility decline, but it may well be sufficient. There can be little opposition to the fundamental reality that adequate management of the economy is an appropriate role for government, and will help reduce fertility as the welfare of the bottom-income groups grows.

CONCLUSIONS

The latter four of the five types of intervention discussed above meet the critical criterion set out: they increase the cost of children to the poor, so that the poor are forced to internalize more fully the social costs of having children, at the same time as they improve the welfare of the poor. They are all win-win policies, in two senses:

• by addressing one or more of the externalities to child-bearing, they make society better off as a whole; the non-poor as well as the poor gain;

• they can be justified anyway, as sound social programmes with high social and economic returns, independent of population concerns.

Whether the consequences of rapid population growth for economic growth and development are large or small, whether the threat to the environment of a growing population is central or tangential, these interventions can be justified—by their high economic and social returns, and their effectiveness in improving the choices and welfare of people. Public policies and programmes to raise human welfare and reduce population growth can be fully consistent, even in the short run. It is a win-win tale.

NOTES

1. The best-known sceptic is Julian Simon (see Simon 1977). A 1986 National Academy of Sciences study (National Research Council 1986) concluded that the empirical evidence linking rapid population growth to slower economic growth in developing countries was surprisingly limited, and that what limited evidence there was suggested that any negative effects were relatively small.

2. An even broader approach would encompass economic reforms which open up economies to domestic and international competition, and reduce public subsidies that penalize agriculture and labour-intensive industries; these tend to penalize the poor, and as discussed later in this chapter, the resulting reduced income of the poor probably induces high fertility.

3. Studies on the determinants of fertility are reviewed in Birdsall (1988). See also World Bank (1984). Birdsall and Griffin (1988) review the evidence that reduced fertility, including among the poor, is associated with lower infant mortality, more-educated mothers, better educational opportunities for children, and better access to family-planning services.

4. Portions of the discussion of population concerns are based on Birdsall and Griffin (1993); and of the discussion of child-bearing externalities on Chomitz and Birdsall (1991).

5. See Simon (1977) for a discussion of positive as well as negative externalities to child-bearing. While not dismissing the possibility of positive externalities in some situations, in this chapter I am concerned with the implications of possible negative externalities.

6. But see Nerlove, Razin, and Sadka (1987) and Willis (1987).

7. An exception is Lee and Miller (1990), who describe a number of externalities related to common property and such public services as education. They conclude (with great caution—they note that the numbers are rough and the exercise is experimental) that externalities are not so great nor always in the expected direction—they are not always negative for developing countries. However, as the authors point out, they do not include any estimate to account for externalities arising from renewable common property resources—the environmental externalities discussed later in this chapter.

8. Evenson (1988) provides estimates of the effect of population on wages and rents in North India. He estimates that a 10% decline in population would boost overall per-capita income by 8%, boost per-capita income of rural landless households by 15%, and reduce land rents by 25%, ignoring scale economies and induced investments associated with population density. If scale economies are taken into account, a reduction in population actually boosts

land rents.

9. The classic presentation of the argument is that of Coale and Hoover (1958). See Birdsall (1988) and Kelley (1988) for reviews of the literature on the economic consequences of rapid population growth in developing countries.

10. See Simon (1977), National Research Council (1986), Birdsall (1988), and Kelley (1988). National Research Council (1986), in turn, cite hundreds of earlier studies and reviews.

11. Lee (1991) sets out this point clearly. He cites that Demeny, in 1972, pointed out that to decry such a decision on parents' part would be like decrying people's decision not to work on Sundays on the grounds that it reduces their incomes.

12. See Romer (1986) and Azariadis and Drazen (1990) on sources of growth including such externalities to human-resource investments.

13. See e.g. Pack and Page (1993).

14. Rosenzweig and Wolpin (1980). This study is useful because parents of twins are not likely to have made a simultaneous or joint decision to have more children and less education, so that lower enrolment rates are more likely to be due to a causal effect of high fertility.

15. See Birdsall and Sabot (1993). They also quantify the effect of larger families on the 'hothouse' effect, i.e. on the extent to which mother's inputs of time per child are affected by the number of children in the home. (Mother's education is also an input to the hothouse effect.)

16. It is important to emphasize that this conclusion goes beyond the usual conclusion that rapid population growth reduces education—to show a negative externality to rapid population growth through its effect on education. It is the negative externality, as will be discussed later in this chapter, that can justify government intervention.

17. See e.g. the report following the 1993 René Dubos Center Conference, dealing with population growth and global warming, land use, pollution, and biodiversity (Revkin 1993). In a recent paper (Birdsall 1993) I compare the costs of reducing greenhouse gas emissions in the United States via a carbon tax to the costs of reducing greenhouse gas emissions in developing countries through family-planning and education programmes that would reduce population growth; I conclude that the latter is as cost-effective as the former and that therefore the developed countries should include as part of any optimal carbon strategy transfers to poor countries to finance family-planning and education programmes to reduce fertility.

18. Cruz (1992) illustrates the former case for the Philippines.

19. The case of the Philippines is a good one. At the same time, it must be said that a good part of forest clearing, particularly of tropical forests in Brazil, Indonesia, and parts of Africa, has not been due to population growth so much

as to road-building and fiscal and pricing policies that made logging commer-
cially attractive.

20. Note that, although wages are depressed in the scenario, rents and profits are
boosted. High fertility is therefore potentially Pareto superior to low fertility,
assuming that there were a mechanism for redistribution of some of the rents
of landlords and capitalists (Ng 1986; Willis 1987).

21. Behrman and Birdsall (1988) show this for males in Brazil.

22. See Naim (1993) for a relatively pessimistic prognosis for consolidation of
reform in Latin America, in part due to the neglect of social issues and the
dire situation of the poor.

23. This is a central message of the report of the World Bank (1993) on East Asia.
See also Birdsall and Sabot (1993) on the 'virtuous circle' of human capital
accumulation, improving income equality, and rising income in East Asia.
For a study of the effects of income distribution on growth at the country
level, see Birdsall, Ross, and Sabot (1994). They attribute the success of East
Asian economies in sustaining high growth in part to their low levels of
income inequality and policies of 'shared growth'.

24. For other evidence on the effect of high fertility on poverty, see Birdsall and
Griffin (1988).

25. Only if a change in their situation makes the poor at least as well off as they
are now should we expect them voluntarily to choose fewer children. Otherwise there would have to be an element of coercion—which would clearly
not increase the poor's welfare.

26. Subsidies or positive incentives to encourage high fertility are more common,
having been used in France and parts of Eastern Europe, and are obviously a
form of incentive that is politically acceptable. See Chomitz and Birdsall
(1991), where the justification for and welfare implications of incentives to
reduce fertility, as well as of child taxes and child quotas, are discussed.

27. Entrapment occurs if an individual is induced, for example, due to myopia or
desperate poverty, to take an irreversible step he or she later regrets. It has
been discussed in the context of cash payments to persons who are sterilized.
They conclude that incentives to correct for failures in the market for information about contraception have potential in many developing countries, but
that incentives to correct for externalities are harder to justify, would probably involve larger financial amounts, and raise a number of difficult ethical
issues.

28. More formally, note that the appearance of a school reduces the relative cost
to parents of ensuring higher 'quality' children versus ensuring a higher
'quantity' of children. Becker and Lewis (1974) present the formal argument
for the interactive effects of changing the price of quality on the price of
quantity and thus on the demand for different relative amounts of these in the

total bundle of 'child services'.

29. It may be counter-intuitive but it is easily demonstrated that interventions that would improve the lives of people, such as education and lower mortality, will increase the 'price' of the quantity of children by raising the cost of inputs required to rear children, such as the opportunity cost of parents' time; see Becker and Lewis (1974).

30. See Crook (1978). In this situation, while initial innovators would encounter social resistance, we would expect a 'tipping' effect (cf. Schelling 1978). Once a critical percentage of families had adopted family planning, the old social norm would be undercut, and the other families would follow suit.

31. A 1984 study of accepters of sterilization in Bangladesh (see Cleland and Mauldin 1987) indicated that, for tubectomy and vasectomy accepters, out-of-pocket transportation and food costs averaged almost twice the average daily male wage. More than one-third of accepters spent more than three times the average daily male wage. More than one-third of accepters eventually faced additional medical complications, and 11% eventually spent a total of almost ten times the average daily male wage. Recuperation time averaged four days for men and ten for women, but was also subject to considerable variation. Some 8% of men and 37% of women needed more than fifteen days to recuperate. Thus, although mean acceptance costs are relatively low, there is a substantial risk of relatively high costs.

32. Family-planning services, if of reasonable quality, have certainly contributed to lower fertility, though they are most effective in settings where women are reasonably well educated. In Bangladesh, where women's education is low, contraceptive use rose from 14 to 31% in the 1980s, and the total fertility rate fell from 6.3 to 4.6. (Ahmed 1987; Griffin 1989).

33. These and other aspects of 'shared growth' are discussed at length in Birdsall and Sabot (1993) and World Bank (1993).

REFERENCES

Ahmed, B. (1987), 'Determinants of Contraceptive Use in Rural Bangladesh: The Demand for Children, Supply of Children, and Costs of Fertility Regulation', *Demography*, 24: 361–73.

Azariadis, C., and Drazen, A. (1990), 'Threshold Externalities in Economic Development', *Quarterly Journal of Economics*.

Becker, G. S., and Lewis, H. G. (1974), 'Interaction between the Quantity and Quality of Children', in T. W. Schultz (ed.), *Economics of the Family* (Chicago).

Behrman, Jere R., and Birdsall, N. (1988), 'The Reward for Good Timing: Cohort Effects and Earning Functions for Brazilian Males', *Review of Economics and Statistics*, 70.

Birdsall, N. (1988), 'Economic Approaches to Population Growth', in H. Chenery and T. N. Srinivasan (eds.), *Handbook of Development Economics*, i (Amsterdam), 477–542.

Birdsall, N. (1992), 'Another Look at Population and Global Warming', World Bank, Policy Research Department, Working Paper No. 1020.

Birdsall, N., and Dixon, J. A. (1991), 'Some Economics of Global Climate Change: The View from the Developing Countries', in *Global Climate Change: The Economic Costs of Mitigation and Adaptation*.

Birdsall, N., and Griffin, C. C. (1988), 'Fertility and Poverty in Developing Countries', *Journal of Policy Modeling*, 10: 29–56.

Birdsall, N., and Griffin, C. C. (1993), 'Population Growth, Externalities, and Poverty' in Michael Lipton and Jaques van deer gaag (eds.), *Including the Poor* (Washington D. C.: World Bank)

Birdsall, N., and Sabot, R. (1993), 'Virtuous Circles: Human Capital, Growth and Equity in East Asia', World Bank, Policy Research Department.

Birdsall, N., Ross, D., and Sabot, R. (1994), 'Inequality and Growth Reconsidered', paper presented at the annual meeting of the American Economic Association, Boston, January 1994.

Bongaarts, J. (1992), 'Population Growth and Global Warming', *Population and Development Review*, 18.

Cassen, R. H. (1993), 'Economic Implications of Demographic Change', *Transactions of the Royal Society of Tropical Medicine and Hygiene*.

Chomitz, K. M., and Birdsall, N. (1991), 'Incentives to Reduce Fertility: Concepts and Policy Issues', *World Bank Economic Review and Research Observer*.

Cleland, J., and Mauldin, W. P. (1987), *Study of Compensation Payments and Family Planning in Bangladesh* (Dhaka: National Institute of Population and Training).

Coale, A. J., and Hoover, E. M. (1958), *Population Growth and Economic Development in Low-Income Countries* (Princeton, NJ).

Crook, N. R. (1978), 'On Social Norms and Fertility Decline', in G. Hawthorne (ed.), *Population and Development* (London).

Crosson, P., and Anderson, J. R. (1991), 'Resources and Global Food Prospects: Supply and Demand for Cereals to 2030', World Bank, Technical Paper No. 184.

Cruz, M. C. J. (1992), *Population Growth, Poverty, and Environment Stress: Frontier Migration in the Philippines and Costa Rica* (Washington, DC: World Resources Institute).

Dasgupta, P. (1993), *An Inquiry into Well-being and Destitution* (Oxford).

Evenson, R. E. (1988), 'Population Growth, Infrastructure, and Real Incomes in North India', in R. D. Lee *et al.* (eds.), *Population, Food, and Rural Develop-*

ment (New York).

Griffin, C. (1989), 'Bangladesh Public Expenditure Review: Public Resources Management During the 4th 5 Year Plan, FY91–95', World Bank, Report 7545.

Kelley, A. C. (1988), 'Economic Consequences of Population Change in the Third World', *Journal of Economic Literature*, 26: 1685–1728.

Lee, R. D. (1991), 'Evaluating Externalities to Child-Bearing in Developing Countries: The Case of India',*World Bank Economic Review and Research Observer.*

Lee, R. D., and Miller, T. (1990), 'Population Growth, Externalities to Childbearing, and Fertility Policy in Developing Countries', *Proceedings of the World Bank Annual Conference on Developing Economics.*

Naim, M. (1993), 'Latin America: Post-Adjustment Blues', *Foreign Policy.*

National Research Council (1986), *Population Growth and Economic Development: Policy Questions* (Washington, DC).

Nerlove, M., Razin, A., and Sadka, E. (1987), *Household and Economy: Welfare Economics of Endogenous Fertility* (New York).

Ng, Y.-K. (1986), 'On the Welfare Economics of Population Control', *Population and Development Review*, 12: 247–66.

Pack, H., and Page, J. (1993),'Accumulation, Exports, and Growth in the High Performing Asian Economies', paper presented at the Carnegie Rochester Conference on Public Policy, April 1993.

Revkin, A. (1993), 'A Bridge between Rio and Cairo', Forum on Population, Environment, and Development, Summary Report, unpublished report prepared for The René Dubos Center for Human Environments, New York City.

Romer, P. M. (1986), 'Increasing Returns and Long-Run Growth', *Journal of Political Economy*, 94: 1002–37.

Rosenzweig, M. R., and Schultz, T. P. (1985), 'The Demand for and Supply of Births: Fertility and its Life-Cycle Consequences', *American Economic Review*, 75: 992–1015.

Rosenzweig, M. R., and Wolpin , K. (1980), 'Testing the Quantity–Quality Fertility Model: The Use of Twins as a Natural Experiment', *Econoamerica*, 48: 227–40.

Schelling, T. C. (1978), *Micromotives and Macrobehaviour* (New York).

Simon, J. L. (1977), *The Economies of Population Growth* (Princeton, NJ).

Westoff, C. F., and Ochoa, L. H. (1991), 'Unmet Need and the Demand for Family Planning', Institute for Resource Development/Macro International, Inc, Demographic and Health Surveys Comparative Studies No. 5.

Willis, R. J. (1987), 'Externalities and Population', in D. G. Johnson and R. D. Lee (eds.), *Population Growth and Economic Development: Issues and Evidence* (Madison, Wis.).

World Bank (1984), *World Development Report: Population Change and Economic Development* (Oxford).

World Bank (1993), *The East Asian Miracle: Economic Growth and Public Policy* (Oxford).

8

Institutional Analysis of Fertility

Geoffrey McNicoll[*]

Few would deny that societies have a legitimate interest in their demographic futures. This is perhaps clearest in the case of immigration, which is subjected to often-stringent control at virtually any national border as an accepted exercise of national sovereignty. For natural increase, the excess of births over deaths, there is potentially an analogous social interest, although the intrinsic decentralization of this case and the countervailing individual interests that are involved render any organized societal response much more difficult.

In poor countries the social interest in demographic change may be compelling. Rapid population growth, the result of greatly reduced mortality but continued fairly high fertility, is a large burden on their development effort. Moreover, both the pace of growth and the resulting size of population have major implications for environmental amenity and ecosystem stability. Indeed, comparable growth in affluent countries, were it to occur, would also pose serious environmental problems, albeit with more scope for remedy. Efforts to moderate fertility levels have a legitimate place in public policy.

Admittedly, the premiss that rapid population growth has these adverse effects is not universally accepted. A recent demurral, for instance, can be found in Göran Ohlin's contribution to the 1992 special issue of *Ambio*: population, Ohlin argues, deserves only a minor place on the agenda of global issues; not much hinges on whether fertility declines quickly or slowly, whether world population reaches a plateau at 10

* Professor, Demography Program, Research School of Social Sciences, Australian National University, Canberra. This is a revised text of a lecture presented in the series on Population, Environment, and Development organized by The Beijer Institute, Royal Swedish Academy of Sciences, Stockholm, on October 14 1993. Comments from seminar participants, and particularly from Partha Dasgupta, Tommy Bengtsson, and Christer Gunnarsson, are acknowledged with appreciation.

billion or at 15 billion. Variants of this view are held by a number of economists, and some demographers. They are expressed in a well-known report of the US National Academy of Sciences (National Research Council 1986).

This is not a debate I wish to enter here. (It is treated at some length in McNicoll 1984. See also Keyfitz 1992.) It does, however, bear on my present topic in one respect. The conclusion that a difference as great as that between the doubling and tripling of world population size, or between 1 per cent and 2 per cent per annum in the population growth rate, can be a matter of unconcern results principally, in my view, from an over-generous estimation of societal adaptability. (Many would argue that there is a similar overestimation of environmental resilience.) In essence, the institutional structure that underpins—indeed, that virtually constitutes—human society is simply neglected. A good part of main-stream economics, and an even larger part of standard demography, have no interest in institutions—or, rather, suppose that institutions adapt as required or can be redesigned at will. Social structures are taken as neu-tral with respect to scale. Resource scarcities are assumed to elicit productivity improvements and shifts to substitutes; many processes of environmental degradation, similarly, have technical and organizational remedies, albeit not always invoked. A fall in the demand for children will occur eventually as a natural consequence of economic develop-ment. Earlier or additional fertility decline, should it be desired, can be achieved through standard-order anti-natalist programmes. Together, the effect of this stance is to break up the triad of population, environment, and development into its components, each of which can be dealt with separately. Whatever feedbacks exist in the larger system work only through effects on the policies of national governments—on their sense of urgency and on consequent legislative and budgetary priorities.

In these pages I shall sketch in part of the missing institutional context of population trends and policies. My brief is to discuss fertility; the obvious focus dictated by present realities is on the processes of fertility transition from high to low levels that can be observed in the contempo-rary world and the strategies that may hasten that transition. This is to revisit a subject I have written on before (McNicoll 1978; 1980), but in the light of an additional decade of demographic experience and recent developments in institutional theory.

1. INSTITUTIONS: A BRIEF INTRODUCTION

Institutions, in the sense used here, do not of course refer to tangible public entities like prisons or hospitals, but to clusters of behavioural rules governing (or, to put it more neutrally, regularities describing) human actions and relationships in recurrent situations. Such rules may be written or unwritten, but are publicly known even if unexpressed; sanctions for violating them may be dictated by external authority or self-imposed. An essential feature of institutions is that they persist, generating a society's distinctive patterns of social organization and the texture of social life.

A useful distinction, due to Carl Menger (1883), is between 'organic' and 'pragmatic' institutions. The first refers to institutions that are inherited from the past, structures that have emerged 'without a common will', such as (to cite Menger's own examples) the law, money, markets, language, the state; the second, to those that are consciously designed as part of efforts to plan, remedy, or reform. (See Schotter 1986: 117–8.) As we shall see, there is less to this distinction than first appears. Institutions evolve and are often much less ancient than supposed—in Eric Hobsbawm's phrase, tradition can be invented; and efforts at institutional design must reckon with the fact that any society is a mass of habits and interests ready to distort or wholly subvert the best intentions of social planners, leaving a veneer of novelty but sometimes little more. '[T]hough governments or nations can in some measure determine what institutions shall be established,' wrote J. S. Mill, 'they cannot arbitrarily determine how these institutions shall work' (Mill 1848: i. 27).

In the familiar opposition between structure and agency, institutions by definition have to do with structure. But they are not hard-cast channels that, once set in place, demand compliant behaviour. They are constantly being made and remade by those coming into contact with them, emerging renewed or marginally changed, or falling into disregard and disuse. The role of agency is distinct, although limited. Thus, acceptance of a structural perspective does not imply a timeless analytical conservatism. Structures are outcomes of continual negotiation and renegotiation, of 'conversation'; they can be 'reinvented' and 'imagined' (Laslett 1979; Hobsbawm and Ranger 1983; Anderson 1991).

Institutions have both material and cultural antecedents. That is, they emerge or are designed as ways for people to deal with recurrent problems about material realities (resources, technologies, environmental conditions, human biology) and as manifestations of ideational systems that give some degree of coherence in the cultural domain of symbol and belief. The economist's conception of an institution as a solution strategy in an iterated game has a loose analogy in the 'game' of seeking (imputing?) meaningfulness in an inchoate or contested world. While it is often an appropriate explanatory tactic to emphasize either material or ideal determination and to reject the other, we can allow that in many situations each is likely to be influential and neither determinative. The consequent analytical problem is the one described by Jane Guyer (1980: 356) for an institutional case that bears closely on reproductive behaviour—the sexual division of labour:

The [sexual] division of labour is, like all fundamental institutions, multifaceted. Within any particular society, it is an integral part of the ideological system, economic organization, daily family life, and often the political structure as well . . . In any one case all these dimensions reinforce each other, so that the current structure seems both heavily overdetermined and ultimately mysterious since it is difficult to assign weight to one factor over another.

Other social institutions connected with fertility plausibly share this multi-faceted nature.

The institutions of economic development, in contrast, are usually accorded straightforward materialist origins. Understanding what promotes or impedes economic growth seems largely to be a matter of understanding the structure of rewards and risks—the institutional setting defined by local economics and politics. It was not always so: the Protestant ethic long held sway in explaining economic development in the West; its more elusive Asian counterpart, the Confucian ethic, still does in some circles. Development policies of 1950s or 1960s vintage that sought to 'make men modern' or teach the 'need for achievement' were strategies premissed on ideational determination of behaviour. These now have a decided air of quaintness about them. Political development, on the other hand, is inescapably in both realms. Joel Migdal (1988: 27) speaks of 'the integration of the material and the moral' in the establishing of systems of social control: tangible rewards and sanctions are combined with the symbolic.

The insight that institutions emerge from behaviour has been the stimulus for an extensive research programme seeking links between individual-level transactions and structural features of societies (see Hechter 1983). The 'new institutional economists' and methodological individualists in sociology have pursued this programme. It faces the problem, however, of settling quite what has to be explained. Social science is always at risk of over-explanation—going too far in imposing coherence and consistency, eliminating happenstance. But institutions are, in a phrase of Clifford Geertz, a precipitate of time. They are replete with contradictions and odd survivals from the past. They adjust (admittedly with some delay, and at times only, so to speak, with considerable provocation) to changing circumstances—say, shifting power relations among social groups or altered transaction costs. But the direction in which they adjust depends not merely on those new circumstances but also, and crucially, on their past history. Social action, Charles Tilly (1993: p. xi) puts it, 'lays down residues that limit the possibilities of subsequent social action'. In a word, institutions are path-dependent.

Path dependency is a simple but powerful idea, one that seems entirely unremarkable to historians but that is generating large changes in the social sciences. Indeed, it is compelling social scientists (and equally, biological scientists) to become historians. As an implication of allowing positive feedbacks in the economy—entailing possibilities for multiple equilibria in place of the unique equilibrium outcome of the standard theory—path dependency has been on the fringes of economic growth theory since the 1920s or earlier. It has broken into the mainstream only with the development of mathematical techniques for describing non-linear systems. In such systems, small perturbations in initial conditions or 'accidental' shocks along the way can lead to large and effectively unpredictable shifts in outcomes. The system can become 'locked in' to one of these alternative equilibrium paths, but which one cannot be foreseen.

In the social sciences these concepts have been developed and applied in the theory of technological change (Arthur 1989; David 1985) and in a number of other areas—particularly, the 'new' trade and growth theories. In the theory of institutional change the arguments are developed by David (1992, 1993) and North (1990). Related but independent institutionalist approaches that apply a somewhat similar range of ideas to economic development, although not concerned with formalization, include those of Mancur Olson (1982) on 'economic sclerosis'—lock-in by rent-seeking interest groups—and E. L. Jones (1988) and Kang Chao (1986) on explaining historical near-misses in industrialization.

Path dependency reminds us that history matters, a phrase fast becoming a cliché in institutionalist writings. History inheres in structure. The institutional setting that exists at any juncture, however coherent and 'logical' it may appear, is in significant measure arbitrary. If the tape were rerun, as Stephen Jay Gould (1989) puts it in reference to the emergence of biological structures, different forms would emerge, no less coherent and logical.

This picture is not meant to be restricted to the stark world of the political realists, with institutional change responding (hesitantly) to the pressures and claims of interest groups and stakeholders in proportion to their accumulated influence and power. As I remarked above, institutionalization also takes place in cultural systems—for example, in ideology. This can similarly be seen as a path-dependent process. Terry Rambo (1991: 91) draws the parallel with evolutionary theory:

[T]he course followed by cultural evolution, like that of biological evolution, is indeterminate. The outcome we observe is the result of a multiplicity of stochastic events. It happened only because all of these events occurred as they did. It would have happened differently if they had not. The search for 'laws' of cultural evolution is thus ultimately futile.

The cultural system imposes its own constraints on how the future is envisaged. It is for that reason that authoritarian regimes see the past as much as the present or future as contested terrain (recall Orwell's Party slogan: 'who controls the past controls the future: who controls the present controls the past'). The creating of myths about the past, of course, is by no means confined to such regimes.

The influence of hopes and expectations about the future on individual decisions and behaviour, and thus, in aggregate, on the processes of institutionalization, distinguishes human institutional change from biological evolution. Expectations are partly endogenous to the society, products of acculturation and the 'social construction' of options and opportunities, but also part-exogenous, formed out of the flood of information, well or poorly interpreted, that immerses virtually every individual in the modern world.

The opposition of history and expectations as influences on structural outcomes is posed by Paul Krugman with reference to industrial location

(1991*a*; 1991*b*). A role for expectations introduces the possibility of bandwagon effects and self-fulfilling prophecy, as individuals, in making their own decisions, take account of how they anticipate others will decide (see Arthur 1992). In some cases, stock-markets for instance, outcomes may rest almost wholly with those anticipatory calculations. In others—very likely in most cases of social change, including fertility change—outcomes derive in some measure from both history and expectations.

For some in a society, the leaders and opinion-makers, hopes and expectations can be translated into deliberate efforts at institutional redesign. I noted earlier the uncertainty attaching to this process, not through want of effort or intent, but because of how old practices and relationships and the interests of existing stakeholders reassert themselves in the new dispensation. In a social system exhibiting multiple equilibria, a small attempted change may be ineffectual—negated by the very elements of the system that give the existing equilibrium position its stability. In the jargon of systems theory, there is a 'basin of attraction' around that position. Large institutional change is achievable in times of crisis or of societal trauma, when the high social and political costs of change are discounted. (These are the circumstances Olson (1982) sees as permitting institutional renewal in sclerotic economies.) There are also occasions in which what might appear a small intervention to a policy-maker—that is, one with low economic and political costs—may set up repercussions with larger consequences. We are inevitably dealing with highly incomplete information about the system's dynamics.

Drawing on the experience of other societies or on nearer-at-hand pilot programmes can reduce this uncertainty. This is a hallmark of intelligent policy-making. The advantage that countries embarking on economic development can supposedly gain from late comer status stems not only from the existence of a shelf of technology from which to pick but also from the store of institutional experience, their own and others', both models of success and cautionary tales of failure. Even regimes that have evinced scant concern over the social costs they impose on their citizenry usually engage in this learning process. The brutal but, in their own terms, successful Chinese land reforms of the early 1950s had their origin in the Communist Party's meticulous experiments in agrarian reform in Yenan during 1935–46; the calamitous efforts at institutional and technological renovation in the Great Leap Forward of 1959 came out of the blue.

2. INSTITUTIONAL ENDOWMENTS AND FERTILITY TRANSITION

How do these ideas illuminate the process of fertility transition, historical and anticipated? Reproductive behaviour, for all its biological content, is bound up in social structure. Fertility transition, whatever else it may be, is an institutional phenomenon. To say that is not to make any judgement about how material and ideational forces interact in bringing about fertility decline. Nor is it to deny a role for individual decision-making in response to individual circumstances, often allowing for straightforward accounts of fertility change in terms of consumer-choice theory. It is the perception of those circumstances and the consequent constraints, largely unseen, placed upon the choice-set, that are at issue. For clarity, I shall first discuss the influences on fertility of inherited and anticipated institutional forms and then the effects of policy interventions.

Since institutions are not neatly classifiable, it is an unprofitable task to attempt a complete enumeration of those that are fertility-related. Any such listing, however, would have to include the institutions giving rise to local patterns of social organization—particularly the family and local community; family and property law and the local dimension of public administration; the stratification system and the mobility paths it accommodates; and the labour market. When we later turn to governmental efforts to influence fertility as a matter of public policy, our interests extend still more broadly to the entire institutional context of state-individual relations.

The characterization of any society by such features reflects its unique history. There is no reason to expect that the fertility regime it exhibits should be identical to that of any other society, or that it will respond similarly to changes in 'inputs' such as new technologies, new resource flows, or, more relevant here, much larger surviving youth cohorts. However, there may well be reason to anticipate some convergence of demographic outcomes, tracing the historical transition to low fertility of the advanced economies. And there may even be reason to expect a long-run convergence of institutional structures and individual behaviours, based on the effect of 'expectations'—in this case, derived from some process of globalization of cultures and technologies.

High fertility in the modern world chiefly characterizes agrarian societies, so I shall concentrate on those in the present discussion. Consider first, by means of some sweeping generalizations (drawing on the analysis given in Cain and McNicoll 1988), how different family systems in such societies respond to the challenges posed by rapid population growth. Characteristics of family systems that are particularly important in determining this response are their control over the establishment of new households, the sex roles they perpetuate, and their old-age security functions.

Rules of family formation in pre-industrial north-west Europe required prior accumulation of property before a person could marry and set up an independent household. In periods of economic hardship saving took longer and marriage was delayed—the famous (to demographers) 'marriage valve'. An equivalent limitation on household establishment existed in Tokugawa Japan, in a society in which preservation of the family landholding and descent line was highly valued. In contrast, the joint family systems found elsewhere in Europe, and still found in much of Asia, sheltered newly married couples within a larger household, cutting the link with the wider economy. The lineage may have provided some analogous shelter in African family systems, although some accounts of the recent evolution of these systems suggest a trend towards self-generated savings as a source of bride-wealth—perhaps the emergence of a kind of African marriage valve.

Sex roles are far more than aspects of the family system—hence the adoption of the term 'gender system' to describe their institutional context. Exploring the maintenance of gender systems entails a broad investigation of economy and society. But social control exercised at the family level is clearly an important part of that maintenance: for example, in setting normative marriage ages, enforcing expectations about familial consent to marriage and resulting property transfers, and determining the acceptability or not of polygyny. The more distinctive the separate roles of husband and wife, the greater the scope for differences in fertility interests, with power relations within the family governing which interests prevail.

Family systems also differ in their function as a welfare provider to elderly dependants. Co-residency patterns in joint family systems offer a promise of old-age security for parents who have surviving children (or sons, in the usual, patrilocal, situation), a promise that becomes more certain as mortality declines. This link between high fertility and old-age security is eventually eroded by diminished expectations that children will in fact supply that support: this is an aspect of the 'wealth-flow reversal' highlighted by J. C. Caldwell (1976). No similar link seems to have existed in the European case—evidenced by the institution of 'retirement contracts' the elderly sometimes entered into with their children.

This is the briefest sketch of what is obviously an intricate topic. However, the implication is clear enough and I think not too controversial: it is that family structures differ in their demographic response to economic pressure, including economic pressure generated by population growth. In particular, both Asian joint-family and African lineage systems appear to have some capacity to insulate marriage and marital fertility from economic conditions.

Family forms, we are told in classic work by John Hajnal (1965, 1982) and by the scholars of the Cambridge Group (Laslett and Wall 1972), persist recognizably intact over the centuries. Plausibly, they are the equilibrium solutions to fundamental problems of organizing reproduction and dealing with child and old-age dependency. How children are protected and socialized, how sexuality is regulated and parenthood affirmed, how claims to property are maintained and transmitted, how the dependent elderly are supported, are questions that in the abstract admit many different answers. (Further questions would need to be added, and the answers are not all independent.) The particular factors that conduce to one answer rather than another may be sociologically quite trivial. Institutionalized, however, each feasible combination defines a family structure. A few among those structures would tend to dominate and exclude others, partly by dint of superior capacity to deliver overall societal 'success', partly by greater innate resilience, partly perhaps by sheer fortuity. Belief systems would develop alongside such structures, generating explicit rules of practice and sanctions for violation. Social groups or statuses 'empowered' by a given structure have evident interests in its maintenance.

The regional distribution of family types, therefore, might be wholly arbitrary. That, in fact, is argued by Emmanuel Todd for family types defined by rules of residence, inheritance, and consanguinity. 'Most family systems exist simultaneously in areas whose climate, relief, geology and economy are completely different. It is impossible to perceive any global coincidence at all between ecological or economic factors and family types' (Todd 1985: 196–7). Admittedly, this is a somewhat extreme position. Jack Goody (1976) argues that agricultural technology exerts some influence over family systems, particularly concerning endogamy. Moreover, the underlying Cambridge Group assertion of prolonged institutional fixity itself does not go unquestioned (see, e.g. Seccombe 1992).

The institutional story is not simply one of disparate family types and their implications for demographic outcomes. The other features of the institutional environment noted above may be just as significant. Thus, the corporate community was an important influence on pre-industrial European and Japanese fertility patterns, as a locus for strong social pressures and sanctions keeping family behaviour in line. In contemporary transitions, somewhat analogous community roles can be seen in China (see Parish and Whyte 1978) and Indonesia (McNicoll and Singarimbun 1983). The absence of comparable community solidarity in much of South Asia, with competing affiliations of kin, class, and caste, has made anti-natalist policy there more difficult (Arthur and McNicoll 1978). The dominance of kinship over territoriality as a settlement principle in most of rural Africa has probably done the same in that region (Hyden 1990).

Community does not connote comity. That quality, not much found in real villages, is not needed for there to be some intra-community pressure regulating demographic behaviour. Of course, even where it was strong, the state and the market economy eventually destroy the local community, usurping its functions and attenuating the relationships among its residents. The option for community mobilization in support of fertility policy, or any other policy, is then lost. Among the usurping structures, the legal system is likely to be especially significant. At one extreme it may offer comparative assurance of contractual and physical security; at the other, a corrupt and partisan regime tainting every transaction. The nearer to the first case, the more stable and predictable the institutional environment and the more that behaviour is likely to conform to a simple consumer-choice model, with fertility levels responding to rising costs of children and the incentive for 'quality–quantity' trade-offs; the nearer to the second, the more the necessity for individuals and families to pursue complex alliance strategies (and perhaps to treat children as hedges against risk) with quite uncertain demographic outcomes.

These local institutional structures—not merely in the 'concrete' sense of community rules of membership and forms of governance, but in the cultural mesh of patterns of social relations and behavioural expectations—tend to persist, for reasons analogous to those applying to family systems. (For some striking evidence of this continuity from Italy, identifying 'the astonishing tensile strength of civic traditions', good and bad, enduring over centuries, see Putnam 1993.) Under high mortality, ties beyond the immediate family are an essential means of risk-spreading, and the reciprocal relationships thus established acquire cultural validity as a 'moral economy'. Depending on the affiliative principle invoked, very different forms of community are generated—perhaps once equally satisfactory modes of social organization, but with different capacities to respond to high rates of population growth. Stability of social organization above the family can also be seen in the pattern of marketing areas— the spatial logic of which yields the hexagons of Christaller–Lösch central place theory—and in systems of social stratification, particularly caste systems.

Close attention to the local organizational context of fertility patterns risks inattention to other institutional dimensions that are likely to be equally relevant in understanding fertility change. Susan Greenhalgh (1990) has set out a number of these in an article exploring what should be the content of a political economy of fertility. She emphasizes the conjuncture of local and global processes: that the sources of local behaviour are found not just in that particular society or even nation, but also in the workings of the international economy and in global communications. She notes the centrality of power relations at all levels of social organization bearing on demographic behaviour, 'from the high politics of international organizations and national bureaucracies lobbying to advance their interests to the humble politics of individual women maneuvering to ensure security in their old age' (p. 95). Class relations and property rights, the staples of explanation of socio-economic change particularly, but not only, in the Marxian tradition, find their place here. For most demographic purposes, power subsumes class, both in overt social relations and in maintaining the cognitive closure that defines a class-specific subculture. As for property rights, it is arguably the security of contract (and of person) associated with a given agrarian system rather than its allocative detail that accounts for most of any demographic impact—a dimension that lies in the province of law and government administration. Allocation has probably mattered in certain common property systems, such as the Mexican *ejido*, where free-rider opportunities could work against collective demographic interests, but this situation seems best analysed in terms of failures of community corporate structure.

In sum, societies have institutional endowments just as they have resource endowments, some combinations of which appear to allow a relatively easy path to low fertility when that is to their economic and social benefit. Other institutional combinations impede and delay that process. The differences between the two may be largely fortuitous in origin. However, institutional structures are not fixed: they evolve, changing the circumstances in which fertility choices are made and thus the choices themselves. As the 'carriers of history' (David 1992), institutions are constrained in how they adapt to the pressures impinging on them, not least from larger youth cohorts. In some measure, moreover, institutions are also emergent manifestations of innumerable individual expectations and idealizations of the future. The resultant directions of institutional change and the scope for public-policy influence on it are the topics I now turn to.

3. DIRECTIONS OF INSTITUTIONAL CHANGE

The historical declines of mortality and fertility in Western countries are broadly linked to economic growth and accompanying social change. These links are the subject of the theory of demographic transition. As a strand of now-disavowed modernization theory, demographic transition theory has seemed to many to have unjustly survived the demise of its progenitor. But rejection did eventually come. The theory was overdetermined, it neglected the subtleties and variability of the transition process, and it abstracted from historical specifics (Greenhalgh 1990: 92). Yet, seen from sufficiently afar, it presented a fair, first-order account of demographic change and its covariates in the large. In its full-blown form (see e.g. Notestein 1964), the theory was at least institutionally rich. Some later variants have sought to strip away the encrustations in an effort to reveal an essential relationship—a risky endeavour, in that there may be no single such relationship at the core.

Historical fertility transition exemplifies this picture, showing similarity across societies at a distance but diversity at close-range. Adding contemporary cases of fertility decline greatly increases the diversity. Yet undeniably there are strong forces making for convergent outcomes as well. Urban industrial life-styles and consumption patterns create similar pressures for low fertility wherever they are found. Ambitions for and anticipations of those futures may have behavioural import much earlier. The degree of institutional convergence to be expected, however, remains controversial.

Take the case of the family. A familiar line of argument associated particularly with Talcott Parsons holds that families increasingly become nuclear in urban industrial societies. Caldwell's transition theory incorporates an assumption of this sort: convergence to the conjugal 'Western' family (Caldwell 1976). Convergence is a plausible expectation for joint families, which after all are nuclear for some part of their life cycle, but may be less so for African systems. The weak conjugality characterizing those systems may as plausibly persist, with the decay of lineage then leading to a Caribbean-type family pattern of visiting unions and largely female-headed households. (See Frank and McNicoll 1987.) Low fertility may be similarly attainable under either family system (the Caribbean region's total fertility rate is now below 3), but the economic and social policies that would best promote it may differ greatly.

Caldwell's theory, of course, is to do with parent–child rather than conjugal relations. The turning-point in fertility transition is posited to be the stage of 'wealth-flow reversal', when children, from being an economic asset to the family, become instead an economic burden. To the extent that children are still wanted, it is for non-economic reasons. That change in the family Caldwell associates with a shift in cultural orientations towards acceptance of nuclear-family models and behaviours purveyed in the media and often in formal education. A more materialist stand is taken by Seccombe (1992), who attributes the reversal to the development of a formal labour market which undermines parental authority exercised through control of property. In either case, the effective change, in so far as parental behaviour is concerned, comes when parental *expectations* about their children are affected. Under the theory's predictions, path-dependent constraints that may formerly have preserved the two kinds of family structures as distinct equilibria are overcome.

An analogous situation may exist at the community level. I remarked earlier that kin-group or factional memberships can be seen as alternative bases to simple residence in defining local-level social organization. The different affiliative principles create alternative equilibrium community structures. For demographic outcomes, the distinction lies in the fluidity of membership in the non-territorial groupings, with the consequent likelihood that no upper limit on size of group would be enforced. Rising population density might eventually bring about the 'peasantization' of such a system, giving significance to village-type communities; or the system's former basis may be preserved, its demographic dysfunctionality notwithstanding. Entrenched local power-holders would tend to support the latter outcome: their ability to capture the reins of programme reform is seen in innumerable failed co-operative movements, community development schemes, and land reforms. Hyden's pessimism about African governance (Hyden 1980) reflects his assessment of the inherent strength of affective ties in competition with nascent civil society.

For many other kinds of social institutions, the equilibrium structure, if any can be detected, is much more complex and possibilities for change more varied. There is, very plausibly, an institutional analogue of 'punctuated equilibrium', with comparatively sudden—and not necessarily policy-driven—shifts in the comparative salience of different structures (see Hodgson 1993). For example, legal systems in some societies are a complex overlay of uncodified traditional law, religious law (perhaps more than one system), and civil law, typically with overlapping jurisdictions in the major fertility-related areas of family and property. Over time, the civil code is likely to emerge gradually as the dominant system; along the way, however, in any particular behavioural domain, the applicable system can change much faster: the alternative is already in place. Less formalized systems of cultural authority are also arenas in the contest for social control, whether in a simple behavioural sense or, more profoundly, in influencing language, categories of thought, and moral values. Benedict Anderson (1991) gives a classic account of how this cultural weaponry has been deployed in the colonial and subsequent nation-building enterprises, using as one example the census-taker's struggle with the concept of ethnic identity. Closer to the fertility domain, one might instance the rapid loss of authority befalling Catholic teaching on birth control in Western countries, in competition with less rigorous situation ethics.

In developing countries, a change that has had large implications for fertility transition has been the expansion of state power at the expense of other social institutions. In communist systems the take-over was virtually complete, as a matter of principle. Even in more pluralist societies, development strategies, influenced by statist philosophies, greatly weakened the independent roles of countervailing institutions of social control. Other social and economic changes, notably widened information sources and the increased mobility of labour, helped to bring about the same outcome. The state thereby gained more power to influence fertility, at the expense of less power able to be wielded by the society. (Migdal 1988 describes this opposition in a more general context.) However, since fertility was hardly a priority item on a crowded development agenda, that power was seldom used.

Recognition of the scope and variety of institutional change is a necessary corrective to the ill-informed notion of institutions as timeless structures. It also greatly complicates the analytical task. Like other kinds of evolutionary change, institutional change is in some degree constrained by the past. Over the long-term, its outcomes may be largely unpredictable, although possibilities may be narrowed by the structure of multiple equilibria and by convergent forces in the society. Among the latter are adjustments to behaviour and expectations: institutionalization is a never-ending social process. In the case of fertility, these complications arise with respect to a raft of institutional arrangements that bear (sometimes quite incidentally) on reproductive behaviour. For a broad picture, fertility transition can be linked to the institutional changes that accompany economic development; however, to understand fertility patterns and the course of fertility change in the detail needed to inform population policy design requires close investigation of those institutional arrangements in the particular society concerned.

4. INSTITUTIONAL ASPECTS OF FERTILITY POLICY

Fertility policies can operate by altering prices and preferences attaching to various behaviours within a given institutional system, or they can seek to modify the system itself. The latter policies are chiefly of interest here. Moreover, policies of either kind rapidly become institutionalized themselves: they can be studied in the same way that other institutions are.

Families are necessarily influenced by a wide variety of government actions, few of which have any such intent. Where that intent does exist, the effectiveness of the action is often unclear. Billboards displaying a beaming couple with two or three well-dressed children have long been part of the publicity component of family-planning programmes; television soap operas conveying the contrasting effects on well-being of small and large families are a more recent popular variant. Quite possibly these have some residual effect, although the construction of a new image of the family is inevitably a much more complex and drawn-out process than establishing a particular consumer-good preference. Family change is also the objective of some legislative efforts: setting minimum ages for marriage; making schooling compulsory and child labour illegal; forbidding a requirement for dowry and enforcing property and inheritance rights for women; and so on. Few states have not enacted these kinds of provisions; they are the stuff of many international instruments and declarations; yet the will or capacity to give effect to them is lacking in many poor, high-fertility countries. Without either an elaborate administrative structure reaching down to the household or a similarly elaborate fiscal system, governments have little purchase on the family economy—or on the demographic behaviour that is influenced by it.

A different route of engagement with the family is the attempt to alter inherent power relations within it. Changes in generational power—of children *vis-à-vis* their parents—is at the centre of Caldwell's view of the influence of mass education on fertility (Caldwell 1980). 'Empowerment' of women is the aim of numerous educational and extension programmes in place or planned by international development agencies. In many societies, it is argued, women desire fewer children than their husbands; giving their preferences more weight would lower fertility. Family-planning extension workers, typically dealing with female clients only, may perhaps have such an empowerment effect: that indeed has been claimed in one report on a well-known experiment on service delivery being conducted in Matlab, Bangladesh (Simmons *et al.* 1988).

Less potentially sensitive for government than interventions at the family level are policies that seek to gain purchase on fertility behaviour through communities or local government. As I suggested earlier, both China and Indonesia seem to have created fairly cohesive corporate communities that could play a role for a time in their respective economic development strategies. In both countries this level of social control was also drawn on in achieving their early successes (those of the 1970s) in fertility decline.

Sustained economic growth and the processes of political development make the opportunity for this policy route short-lived. A major ingredient of that growth would often have been grass-roots economic freedom, setting up the dynamics of capitalist development. (Recall the argument of Stephen Cheung (1982) on the irreversibility of change following from land privatization in China after the first Dengist reforms.) Expansion of the labour market reduces the significance of community boundaries. Authoritarian rule at the local level is undermined, but so too are the local institutional bases of community solidarity. A more private and autonomous domestic sphere takes on a new inviolacy and the scope for fertility policy is radically reduced. Fortunately, that same economic growth is also likely to have lessened any need for such policy.

For many societies, of course, the policy route vanished before it could be tried, a casualty of the seeming imperative for centralization of government authority. Efforts to extend government progammes into localities are still made in those circumstances, but are successful only to the extent that there is a demand for the programme's services at the price offered. (How much demand there is for fertility regulation services and the shape of the demand curve are both highly contentious issues—a remarkable situation after twenty-five years or more of programme activity.) For some other societies a comparable route could still be constructed: not mimicking the earlier examples but making innovative use of each society's own institutional inheritance.

Fertility policy that is designed with close reference to the institutional setting in which it is to operate should be able to marshall collateral support and lessen the social costs it imposes (or their equivalent political costs). But if the policy set from which the choice is made is itself truncated, any selection may be far from optimal in that respect. Such a truncation may well have taken place. In the earlier discussion of institutional change, path dependency entered in understanding the variation of fertility regimes and the constraints on their directions of change. Path dependency is also relevant in explaining how fertility *policy* has evolved.

It is a striking, though seldom-noted, fact that the strategy for fertility reduction that Western countries promote for the Third World is unlike anything that could be drawn from their own experience. That experience, as I have remarked, included, for Europe and Japan, an initial rural setting characterized by firm, community-based regulation of marriage and of establishment of a household. There and in the United States fertility transition, when it came, was associated with a series of social and economic changes that strongly pushed up the perceived costs of children to parents: mass education—but typically in good measure locally financed; less child labour and more access by women to the formal labour market; fewer kin over which to share child-raising costs in a more mobile and urbanizing society; new consumption opportunities; and so on. Accompanying these were cultural changes such as the romanticization of childhood and new-found moral obligations of parenthood. (Governments, to the extent that they took any interest in fertility, sought to preserve familistic values against the flood, ultimately in vain.) Yet what these countries recommended to others were government extension schemes to provide subsidized contraceptive supplies and services—schemes at odds with the increasingly anti-statist views found in other areas of development policy.

A partial explanation as to why this happened, in the spirit of the earlier discussion, would be accidents of history. For its nineteenth- and early twentieth-century advocates, such as Marie Stopes and Margaret Sanger, contraception was a means of emancipating and empowering women. Sanger in particular was also alarmed over the problems of global population growth—evident in her successful efforts to mobilize scholars that led to the formation in the 1920s of the International Union for the Scientific Study of Population and the Population Association of America. To her, the natural answer to rapid population growth was the neo-Malthusian one: a world-wide acceptance of contraception. The policy content of classical demographic transition theory was minimal—reducible, in a later formulation, to the slogan that development is the best contraceptive. But when, in the 1950s and 1960s, a practical intervention to restrain fertility was sought, instead of thinking harder about the micro-institutional basis of past transitions, the family-planning model was at hand. Development theory at the time emphasized government direction, so the public sector was granted a virtual monopoly. The medicalization required by some modern methods of birth control gave programme authority to a strong professional group, the doctors. Foreign-aid and international agencies became powerful advocates of family-planning programmes, with research funds directed accordingly. (For the United States, the story from transition theory to family-planning programme in the context of foreign policy and the Cold War is told compellingly by Simon Szreter 1993; on inward-looking funding priorities, see Demeny 1988.)

At a finer-grained level, the choice of the technology of birth control has also shown signs of path-dependent development. The case of the persistence of 'traditional' and quite ineffective contraceptive methods in modern-day Japan, backed up by abortion, is attributed by Samuel Coleman (1983) to a powerful medical lobby protecting its turf (medical abortion) against threat (oral contraception, sterilization), to a private family-planning association that gains its income from sales of condoms, and to entrenched sex-segregated conjugal roles within the Japanese family. Greenhalgh (1990: 98) remarks of the case: 'In Japan, and probably elsewhere as well, contraceptive practice is embedded in relationships of power and economic advantage at every level of the family-planning system, from national bureaucracy, to local family-planning association, to individual couple.' In India, one can surmise, the fact that sterilization is the most prevalent contraceptive method is not a conseqence of some cultural quirk or of economic exigency, but is a similar case of technological, administrative, and political lock-in.

Institutional dominance creates and then is reinforced by dominance of ideas. That a country seeking to lower its fertility should establish a family-planning programme now has the status of a truism. China, where fertility policy was in many respects an entirely different process, is added to the collection of success stories—typified as a strong and effective family-planning programme, albeit with an unfortunate coercive tinge. Indonesia, also with some amount of muscle applied in the early years, is seen as entirely mainstream. But this lumping together has the unfortunate result of obscuring such distinctions, creating the impression—and often the reality—that a family-planning programme is something that can be taken off the shelf. A more illuminating classification of anti-natalist policies would be in terms of the governance and local organizational structure of society, with the means of contraception and ways of delivery secondary issues.

5. VARIETIES OF FERTILITY TRANSITION

The perspective on fertility I have set out in this chapter can be illustrated by a discussion of recent world experience of fertility trends. I am not concerned with comprehensive description, but with a brief, selective presentation of differences in structure and outcome. My starting-point is a fourfold typology of broad patterns of fertility transition, roughly identified with world regions. These might be telegraphically referred to as 'traditional capitalist' (Latin America), 'soft state' (South Asia), 'radical devolution' (China), and 'growth with equity' (East and South-East Asia apart from China). A fifth pattern, as yet showing only slight signs of fertility decline, could be termed 'lineage dominance' (Sub-Saharan Africa). I borrow these titles and parts of the following descriptions from an unpublished paper by Paul Demeny and myself. The demographic data are from United Nations (1991).

Traditional capitalist. The modern variant of the kind of fertility transition experienced by Europe and North America is probably best represented by the large countries of Latin America outside the so-called southern cone (where fertility has been moderately low for decades), notably Brazil and Mexico. Here total fertility has fallen by half, from 6 to 3, since the early 1960s.

These trends seem tied to socio-economic changes of the kind that enter classic statements of demographic transition theory. Mortality had declined substantially by the end of the 1960s in each of these two countries, and the 1970s saw rapid urbanization and falling shares of the labour force in agriculture. By 1980 two-thirds of the population was urban. Per-capita incomes were high by Asian or African standards, but the averages obscured large inequalities—and falls as well as rises over time. For both rich and poor, however, high fertility became increasingly incompatible with urban industrial life.

Urbanization, and the fall in private demand for children that accompanies it—in *favela* as much as in suburb—is clearly a substantial part of this story. It is also an uninteresting part: low urban fertility is neither a surprise nor a puzzle. The (smaller) rural decline may owe something to the conventional demand-effects of income and educational advances, but possibly more to the agrarian institutional structure. In the Brazilian case, consolidation of land into larger holdings led to the mass conversion of tenant farmers into wage labourers—a process of proletarianization—with likely dampening effect on fertility. (The same process led to massive rural–urban migration.) In Mexico the labour absorption capacities of the rural sector, particularly through the *ejido* system, had plausibly delayed a fertility decline until the system became demographically saturated.

Government concern and involvement with family-size decisions and promotion of contraception were quite strong although late in coming in Mexico. In Brazil they were largely absent: indeed, earlier energetic public efforts to introduce modern contraception could very likely have hastened the pace of decline, lessening the high social costs of the route actually followed.

Soft state. A somewhat different model of fertility transition is epitomized by much of the Indian subcontinent. Along a broad northern swathe of this region, encompassing Pakistan, Uttar Pradesh, Madhya Pradesh, Bihar, and Bangladesh, fertility decline has been slow and, for the most part, fairly slight. In many other Indian states and in Sri Lanka fertility has dropped significantly. Overall, India's fertility has fallen from about 6 to about 4 over four decades. United Nations projections, even though they assume rapid fertility decline to replacement level in South Asia, point to a 70 per cent population increase over the next thirty years in that region, in contrast to 28 per cent for East Asia. India is expected to overtake China in population by about 2030.

In the terms of the earlier discussion of institutional endowments, South Asian social structure appears to have few homeostatic qualities— showing instead a picture of caste- and class-riven villages and the lower reaches of government captured by local élites. (Establishment of more accountable local administration—accountable both to higher administrative tiers and to the citizenry—appears to have been a major factor in the successes in social development seen in Kerala and some other Indian states.) Families, however, may well have experienced a Caldwell-style shift toward nuclearized autonomy, and with it a more exposed fertility calculus. At the same time, an array of cultural and economic changes have probably reduced parental expectations of old-age support from their children.

The role of the state in the development process, influenced by the administrative inheritance from colonial times and the emphasis on central planning that was then popular in development theorizing, has been ambiguous: strong, often meddling, but in important respects undemanding. The epithet of 'soft state', bestowed on countries of South Asia by Gunnar Myrdal, referred specifically to government inability or reluctance to impose obligations on people—not a bad thing in many respects, unless (as often was the case) this simply left the way clear for extra-governmental exploitation or *sub-rosa* bureaucratic rent-seeking.

Family planning conforms to this picture. A family-planning programme was begun in India in the early 1950s and under Health Ministry auspices took on the heavily bureaucratized features of other Indian government activities. Programme targets and performance statistics had to be conveyed through a multi-layered hierarchy able to diffuse accountability and inclined to exaggerate results. Large inter-state differences in programme effectiveness emerged, reflecting distinctive administrative cultures and local socio-economic conditions. While monetary incentives to clients have played a long-standing role in the programme, aside from a short period during the Emergency of 1975–6 force and administrative pressure have been generally absent. Pakistan and Bangladesh have similarly well-established family-planning programmes with either negligible (Pakistan) or modest (Bangladesh) success records.

There is much in the South Asian fertility record that resembles the 'traditional capitalist' model. Moreover, a principal difference from that model as it applies to the broader economy—namely, an interfering government and large public sector—is now being narrowed as deregulatory policies take effect. Although governments of the region, especially India's, have East Asian patterns in mind as they chart new development paths, the outcome, at least for fertility, may be more like Brazil, with the impetus for fertility decline being the changing private demand for children in an institutionally 'thin' environment.

Radical devolution. The most distinctive model of fertility transition in world experience is the case of China, where the total fertility rate plummeted from 6 children per woman in the 1960s to below 2.5 by the end of the 1970s. While having no resemblance to the capitalist path, this transformation was similarly achieved at high social cost.

The conditions that made possible this dramatic change were set up in the massive agrarian reforms of the 1950s—not the short-lived, abortive efforts late in the decade to establish communes as the basic unit of rural society, but the earlier land reform and lower-level collectivization that built on and mobilized pre-existing neighbourhood and village solidarities while destroying the landlord and lineage interests that had overlain them. Local territorial solidarity was reinforced by collective agricultural taxes and the obligations imposed on villages to fund many of their own social services. The boundaries of local units were further accentuated by government restraints on migration and economic diversification. Under these circumstances it would be reasonable to expect that high-fertility incentives within these units would rapidly erode.

Complementing this radical devolution of economic and social accounting was the strong political and administrative structure established by party and government that could convey programme directives from the centre and apply rewards and sanctions for performance. This structure proved highly effective for delivery of health services, with life expectancy reaching 65 years by the late 1970s, well above the levels that experience elsewhere would predict to correspond with China's economic status at that time. When state policy shifted to anti-natalism in the 1970s, it was equally effective with family-planning services. The decline was speeded by political pressures, backed by sanctions, applied by local cadres, transmitting provincial and national dictates. But it is unlikely that the government could have prevailed to the degree it did had there been intense public resistance to smaller families.

The system that yielded considerable success in meeting government health and family-planning objectives was much less effective in generating dynamic economic performance. The Dengist reforms after 1979, effectively restoring private incentives (and obligations: the 1980 marriage law made support of elderly parents primarily a family responsibility), released the family-level economic dynamism that had been blocked by the former collective arrangements and by political and bureaucratic dead weight. Maintenance of stringent control over fertility, even extending to the attempted limitation to one child per family, has proved increasingly difficult as the economy burgeons under 'market Leninism' and as a large floating population grows up beyond official surveillance. However, China's status as a low-fertility country, with family-size choices for the great majority of couples securely contained in the range of one to three children, is not likely to be affected.

Growth with equity. The outstanding success of the East Asian hyper-growth economies, led by Japan, that has made that region such a large contributor to world manufacturing trade, and Taiwan and South Korea exemplars for development policy, has its counterpart on the demographic side. Fertility in Taiwan started to fall in the early 1950s, in South Korea a decade later; both are now at levels below replacement. The so-called second-tier countries of South-East Asia, especially Thailand, Malaysia, and Indonesia, have also shown strong economic progress and substantial fertility declines. (The Philippines, its political development in some respects closer to Latin American models, lags on both scores.)

The standard description of the Korea–Taiwan model is 'growth with equity', contrasting with the former orthodoxy of development theory that saw redistribution as in conflict with growth. Fairly radical land reforms in the 1950s laid the basis for small-scale peasant production, supported by effective government extension and financial services. Education and health similarly received high government priority and, gradually from the 1960s, so did family planning. Impressively, the shift to an industrial economy was accomplished without the price distortions, large urban–rural wage differentials, and growing income inequality that typically attend these processes.

In contrast to the 'traditional-capitalist' model, government roles in this case extended well beyond the vigorous promotion of state and private economic enterprise to cover also the enlightened provision or support of a variety of social services. And, in contrast to the 'soft-state' case, in place of a fairly rough-and-ready, class-riven rule of law, there was stringent, even meticulous, administrative and political control. Family-planning programmes, as these developed, however, were and remained defined for the most part as a social service rather than an aspect of government development administration. This was in sharp distinction with China: these governments were authoritarian where they chose to be, but conceded a wider domain of individual privacy surrounding family life.

Fertility began to drop in Taiwan and South Korea before family-planning programmes took hold. Most observers would agree that the programmes facilitated and, to some degree and among some groups, speeded the change. But the underlying course of demographic transition probably had more to do with the style and pace of economic and cultural change than with the details of programme design and operation.

Family-planning programmes are plausibly accorded a greater role in the fertility declines of the second-tier countries, notably Thailand and Indonesia. These two countries, together perhaps with Mexico and Colombia, are exhibited as the chief examples of family-planning pro-

gramme success. As with Mexico and Colombia, however, it is the
timing rather than the fact of fertility decline that has probably been
influenced. In Thailand a growing demand for contraceptive services,
stimulated by a fast-growing consumer economy, would likely have elic-
ited development of a private-sector supply system: there were no
apparent cultural obstacles in the way of demand increase. In Indonesia
the programme's initial successes owed much to the coincident creation
of a much-strengthened local government administrative system through
which policy objectives could be transmitted and local officials held
accountable for results. Service demand subsequently caught up with this
top-down programme as families confronted the realities of how best to
assure their children access to secondary education and modern-sector
employment. Moves towards a user-pays programme are now in train.

Lineage dominance. For the most part the course of fertility in other
regions of the world resembles or seems likely to resemble one or other
of these four models. The principal possible exception is Sub-Saharan
Africa. Over coming years this region may fall into line—say, into a kind
of South Asian variant: that is the more-or-less explicit expectation of the
major international agencies. The UN estimates Sub-Saharan Africa's
total fertility to have been virtually unchanged, at a level of around 6.6,
from 1950 to 1990 (more recent survey data from a few countries, notably Kenya, show fertility starting to fall in the 1980s); the medium-
variant projections then indicate fertility dropping to 5 by 2010 and 3 by
2025. But, as the earlier discussion of this situation implied, it is at least
as plausible that Africa will trace a demographic path of its own, one
reflecting its distinctive family systems and its kin-based, rather than
territorial, pattern of social organization and governance.

Rapid population growth has meant that increasing parts of Africa are
shifting from conditions of labour scarcity to land scarcity. The resulting
economic pressures toward privatization of land and lessened demand
for rural labour should be expected to have anti-natalist effects. The
residue of lineage influence—over land and over family behaviour, particularly marriage—remains powerful in parts of the region, tending to
counter those effects, but the trend toward a falling demand for children
will no doubt soon dominate if it has not already begun to do so. A more
uncertain factor with a large potential to influence the region's fertility
future is the AIDS epidemic, both in its immediate demographic impact
and through its deeper consequences on economy and culture.

Condensed description and speculation are no substitute for detailed analysis, although the summaries I have just given certainly draw on much fuller accounts. But they can point to features of social setting and policy that are salient for demographic outcomes. In this case, one might observe the manner in which successful policies mobilize and work through the existing local social organization; the consequent need for attention to timing and sequencing of interventions, as the development process alters those local structures; the limited scope for but potentially crucial effect of 'getting institutions right'—the often neglected analogue of the World Bank dictum of 'getting prices right'; and the variant social costs imposed by different policies, or amounting to the same thing the variant political capital expended.

6. GLOBAL FERTILITY PROSPECTS AND THE CAIRO AGENDA

In 1960 the global average level of total fertility (average lifetime births per woman) was about 5.0; by 1990 it had fallen below 3.5. If replacement level fertility (2.1 under fairly low-mortality conditions) is taken as a 'goal'—promising, it may be hoped, an eventual zero rate of population growth—then we are just over half-way there.

The second half of the fertility transition to replacement level is expected by the UN forecasters, so far remarkably prescient, to take the next forty to fifty years. During that same period, if they are correct, population growth will add some 4 billion people over the present 5.5 billion. Subsequent increases, reflecting momentum effects, will add another 2 billion, giving a global population levelling out at about 11 or 12 billion by the end of the next century.

Being suitably cautious, given the high sensitivity of population size to very small changes in fertility, the forecasters bracket this medium-variant trajectory by low and high variants, reflecting faster and slower fertility declines. The corresponding populations for 2100 are 6 and 19 billion. Empirical studies of success rates of other projection exercises suggest that these values define a confidence interval of two-thirds.

Most demographers, indeed probably most social scientists, accept the medium-variant path as a plausible future. Fertility will continue on its downward course, under the existing range of policies, until flattening out (for reasons of neatness?) at replacement level. Somehow the world will learn to live with a final doubling of the population, assisted by technological developments as yet unforeseen. After all, it is often pointed out, fifty years ago when the world's population was little over 2 billion, most people would have been alarmed at the prospect of accommodating today's numbers. Aesthetic losses are put out of mind and eventually forgotten: they are irrelevant could-have-beens. The more tangible limits to growth have had a tendency to retreat as they are neared, like the limits on London's population once presumed to be set by the stench and disposal problems of horse manure. If so, my subject of fertility transition and its determinants may soon lose much of its remaining public interest. It would be left to the social historians, while the world moved on to other business. Although I would find that an unsatisfactory ending, it is probably the way that most public policy issues are eventually dealt with.

But there is much to go wrong. We are all aware of the signs of environmental stress associated with rapid population growth and land scarcity in agrarian societies, and with the vulnerability of the vast urban populations that are building up. More recently we have come to realize that there are global environmental changes in prospect, with potentially drastic implications for agricultural growth and ecological stability. The casual expectation that human health would continually improve up to some biological limit is having to be discarded, as new threats of infectious disease emerge. It is entirely possible that fertility levels will not follow their assigned downward path.

Exploring the processes of fertility transition is not an academic byway but a central requirement for understanding the population dimension of these problems and assessing policy responses to them. For about two decades after the second World War it was taken as understood that demographic transition was directly tied to economic development. Poor countries would successively follow the routes of the West, and later Japan, not only to industrial affluence but also, largely in consequence, to low mortality and fertility. A Bucharest-style *laissez-faire* fertility policy was implied. Starting in the 1960s, an even simpler view has gained acceptance, proclaimed at the Mexico City Conference of 1984: that fertility has responded to public family-planning programmes promoting the advantages of smaller families and proffering subsidized contraceptive supplies and clinical services. The conviction that we know what to do but just need more resources to do it is a prevalent sentiment in international agency circles and common enough among demographers; it will be a prominent assertion at the 1994 Cairo Conference.

I have argued here for a third view, derived from an institutionalist interpretation of fertility transition. It is, I believe, better founded in social scientific terms than the others. The policy directions it calls for are not laid out in capsulated general theory or in an imported programmatic recipe. They will emerge from thorough, situated analysis of the institutional setting of fertility, sceptical assessment of policy experience elsewhere, innovative design to mesh with and, where feasible, to enlist support from existing or newly modified institutional structures, and careful experimentation. The problem of high fertility is not that fertility will never decline: assuredly it will. The problem is to find routes to low fertility consistent with high achievement in economic and social development and limited sacrifice of environmental values. It would be surprising if such routes did not entail close institutional engagement.

REFERENCES

Anderson, B. (1991), *Imagined Communities: Reflections on the Origin and Spread of Nationalism*, rev. edn. (London).

Arthur, W. B. (1989), 'Competing Technologies, Increasing Returns, and Lock-in by Historical Events', *Economic Journal*, 99: 116–31.

Arthur, W. B. (1992), 'On Learning and Adaptation in the Economy', Santa Fe Institute, Santa Fe, New Mexico, Working Paper, 92–07–038.

Arthur, W. B., and McNicoll, G. (1978), 'An Analytical Study of Population and Development in Bangladesh', *Population and Development Review*, 4: 23–80.

Cain, M., and McNicoll, G. (1988), 'Population Growth and Agrarian Outcomes', in R. D. Lee *et al.* (eds.), *Population, Food and Rural Development* (Oxford).

Caldwell, J. C. (1976), 'Toward a Restatement of Demographic Transition Theory', *Population and Development Review*, 2: 321–66.

Caldwell, J. C. (1980), 'Mass Education as a Determinant of the Timing of Fertility Decline', *Population and Development Review*, 6: 225–55.

Chao, K. (1986), *Man and Land in Chinese History: An Economic Analysis* (Stanford).

Cheung, S. N. S. (1982), *Will China Go 'Capitalist'? An Economic Analysis of Property Rights and Institutional Change* (London).

Coleman, S. (1983), *Family Planning in Japanese Society: Traditional Birth Control in a Modern Urban Culture* (Princeton, NJ).

David, P. A. (1985), 'Clio and the Economics of QWERTY', *American Economic Review*, Supplements, 75: 332–7.

David, P. A. (1992), 'Why are Institutions the "Carriers of History"?', paper presented at the Stanford Institute for Theoretical Economics.

David, P. A. (1993), 'Historical Economics in the Long Run: Some Implications of Path-Dependence', in G. D. Snooks (ed.), *Historical Analysis in Economics* (London).

Demeny, P. (1988), 'Social Science and Population Policy', *Population and Development Review*, 14: 451–79.

Frank, O., and McNicoll, G. (1987), 'An Interpretation of Fertility and Population Policy in Kenya', *Population and Development Review*, 13: 209–43.

Goody, J. (1976), *Production and Reproduction* (Cambridge).

Gould, S. J. (1989), *Wonderful Life* (New York).

Greenhalgh, S. (1990), 'Toward a Political Economy of Fertility', *Population and Development Review*, 16: 85–106.

Guyer, J. (1980), 'Food, Cocoa, and the Division of Labour by Sex in Two West African Societies', *Comparative Studies in Society and History*, 22: 355.

Hajnal, J. (1965), 'European Marriage Patterns in Perspective', in D. V. Glass and D. E. C. Eversley (eds.), *Population in History* (London).

Hajnal, J. (1982), 'Two Kinds of Preindustrial Household Formation System', *Population and Development Review*, 8: 449–94.

Hechter, M. (ed.) (1983), *Microfoundations of Macrosociology* (Philadelphia, Pa.).

Hobsbawm, E., and Ranger, T. (1983) (eds.), *The Invention of Tradition* (Cambridge).

Hodgson, G. (1993), *Economics and Evolution: Bringing Life Back into Economics* (Oxford).

Hyden, G. (1980), *Beyond Ujamaa in Tanzania: Underdevelopment and an Uncaptured Peasantry* (Berkeley, Calif.).

Hyden, G. (1990), 'Local Governance and Economic-Demographic Transition in Rural Africa', in G. McNicoll and M. Cain (eds.), *Rural Development and Population: Institutions and Policy* (New York).

Jones, E. L. (1988), *Growth Recurring: Economic Change in World History* (Oxford).

Keyfitz, N. (1992), 'Seven Ways of Causing the Less Developed Countries' Population Problem to Disappear—In Theory', *European Journal of Population*, 8: 149–67.

Krugman, P. (1991*a*) *Geography and Trade* (Cambridge, Mass.).

Krugman, P. (1991*b*) 'History versus Expectations', *Quarterly Journal of Econo-*

mics, 106: 651–67.

Laslett, P. (1979), 'The Conversation between the Generations', in P. Laslett and J. Fishkin (eds.), *Philosophy, Politics and Society*, 5th ser. (New Haven, Conn.).

Laslett, P., and Wall, R. (1972) (eds.), *Household and Family in Past Time* (Cambridge).

McNicoll, G. (1978), 'Population and Development: Outlines for a Structuralist Approach', *Journal of Development Studies*, 14: 79–99.

McNicoll, G. (1980), 'Institutional Determinants of Fertility Change', *Population and Development Review*, 6: 441–62.

McNicoll, G. (1984), 'Consequences of Rapid Population Growth: An Overview and Assessment', *Population and Development Review*, 10: 177–240.

McNicoll, G., and Singarimbun, M. (1983), *Fertility Decline in Indonesia: Analysis and Interpretation* (Washington, DC).

Menger, C. (1883) *Problems of Economics and Sociology*, trans. Francis J. Nock (Urbana, Ill., 1963).

Migdal, J. S. (1988), *Strong Societies and Weak States: State–Society Relations and State Capabilities in the Third World* (Princeton, NJ).

Mill, J. S. (1848), *Principles of Political Economy* (Boston).

National Research Council (1986), *Population Growth and Economic Development: Policy Questions* (Washington, DC).

North, D. C. (1990), *Institutions, Institutional Change and Economic Performance* (Cambridge).

Notestein, F. (1964), 'Population Growth and Economic Development', *Population and Development Review*, 9(1983): 345–60.

Ohlin, G. (1992), 'The Population Concern', *Ambio*, 21: 6–9.

Olson, M. (1982), *The Rise and Decline of Nations: Economic Growth, Stagflation, and Social Rigidities* (New Haven, Conn.).

Orwell, G. (1949), *Nineteen Eighty-Four* (London).

Parish, W. L., and Whyte, M. K. (1978), *Village and Family in Contemporary China* (Chicago).

Putnam, R. D. (1993), *Making Democracy Work: Civic Traditions in Modern Italy* (Princeton, NJ).

Rambo, A. T. (1991), 'The Study of Cultural Evolution', in A. T. Rambo and K. Gillogly (eds.), *Profiles in Cultural Evolution* (Ann Arbor, Mich.).

Schotter, A. (1986), 'The Evolution of Rules', in R. N. Langlois (ed.), *Economics as a Process: Essays in the New Institutional Economics* (Cambridge).

Seccombe, W. (1992), *A Millennium of Family Change: Feudalism to Capitalism in Northwestern Europe* (London).

Simmons, R., Baqee, L., Koenig, M. E., and Phillips, J. F. (1988), 'Beyond Supply: The Importance of Female Family Planning Workers in Rural Bangladesh', *Studies in Family Planning*, 19: 29–38.

Szreter, S. (1993), 'Policy and Social Science: The Idea of Demographic Transition and the Study of Fertility Change', *Population and Development Review*, 19.

Tilly, C. (1993), 'Series Editor's Preface', in C. Lloyd, *The Structures of History* (Oxford).

Todd, E. (1985), *The Explanation of Ideology: Family Structures and Social Systems*, trans. D. Garrioch (Oxford).

United Nations (1991), *World Population Prospects* (1990) (New York).

9

The Relevance of Malthus for the Study of Mortality Today: Long-Run Influences on Health, Mortality, Labour Force Participation, and Population Growth

Robert Fogel[*]

Research during the past two decades has produced significant advances in the description and explanation of the secular decline in mortality. My assigned task in this chapter is to reconsider Malthus's theory of mortality on the basis of these findings and to assess what aspects of his theory might remain relevant to policy-makers in both rich and poor countries.

In the course of this chapter I will argue that the secular decline in mortality, which began during the eighteenth century, is still in progress and will probably continue for another century or more. The evolutionary perspective presented in this chapter focuses not only on the environment, which from the standpoint of human health and prosperity has become much more favourable than it was in Malthus's time, but also on changes in human physiology over the past three centuries which affect both economic and biomedical processes. A great deal of emphasis is placed on the interconnectedness of events and processes over the life cycle and, by implication, between generations.

Such a perspective conflicts with much of Malthus's theory of morta-

* Walgreen Professor of American Institutions, University of Chicago. Prepared for presentation to the Population-Environment-Development seminar, The Royal Swedish Academy of Sciences, 13 September 1993. This chapter draws on materials in my forthcoming book (Fogel 1995) and on several papers (Fogel 1987, 1991, 1992, 1993, 1994; Fogel and Floud 1994; Fogel, Costa, and Kim 1994). I have benefited from the insights and criticisms of Christopher J. Acito, Dora L. Costa, and John M. Kim. The research reported here was supported by grants from NIA (#5-PO1-AG10120-02), NSF (#SES-9114981), the Walgreen Foundation, and the University of Chicago.

lity. In his view 'normal' life expectation was fixed, and society was evolving toward a stationary state. Variations in mortality were related to the pressure of population on the land and on the existing stock of capital. Although Malthus allowed for the possibility that the arts might change in a salutary direction, the room for such improvements received scant attention in all the various editions of *An Essay on the Principle of Population* (1798). There is not the slightest inkling that Malthus foresaw a Britain with four times the population of his day and fifty times its income. As far as longevity is concerned, Malthus's best hope was that through prudent restraint the poor might approach the health and life expectations of the rich, which at the close of the eighteenth century was less than fifty years at birth. He certainly did not foresee that even the imprudent poor, living in a world of vice that rivalled anything he preached against, would in the 1990s have life expectations half as large again as that which he considered the upper limit of life.

Despite all the different ways in which Malthus misread the future, the central proposition of *An Essay*—that much misery was caused by the pressure of population against available resources—remains valid today, particularly for the poor countries of the world. Although the world is much richer than it was two centuries ago, it is still not rich enough to provide all of the poor nations of the world with the standard of life that has become conventional in the OECD nations. And the degradation of the environment that is associated with the unprecedented increase in population since the Second World War threatens not only to impede progress in the Third World, but to impede, if not reverse, the progress of the OECD world.

This chapter is divided into three parts. Section 1 deals with recent findings about the relationship of nutrition and physiology to the secular decline in mortality in Europe and America, and the bearing of these findings on Malthus's theory of mortality. Section 2 reconsiders the implications of Malthusian theory for economic and health policies in the Third World in the light of recent findings on secular trends in health and certain aspects of human physiology in Europe and America. Section 3 is concerned with a set of issues that are outside the Malthusian intellectual framework and that are likely to become predominant over the next several decades. Central to the new work are the debates on the upper limit to life, on the capacity to roll back the onset of chronic diseases through alterations in life-style and medical interventions, on the possibilities for extending economic productivity past age 65, and on the implications of these issues for meeting pension obligations during the next three to seven decades.

1. SOME NEW FINDINGS ON THE SECULAR DECLINE IN MORTALITY

During the middle of the eighteenth century mortality rates began to decline in England and France.[1] These declines continued for about three-quarters of a century and then levelled off for about a half-century. The declines resumed during the last quarter of the nineteenth century and have continued to the present. Similar movements in death rates have been reported for the Scandinavian countries and for North America. However, it was not until the second decade of the twentieth century that demographers and epidemiologists became aware that the West was in the midst of a major secular reduction in mortality, and at first only the most recent wave of the decline was apparent. A concerted drive to explain the trend began in the 1920s, one aspect of which was an effort to develop a series that would not only show how far back in time the decline extended but would also show its spatial dimensions. Work on documenting the earlier phase of the decline began near the beginning of the first World War and continues to the present.

Between the late 1930s and the end of the 1960s a consensus emerged on the explanation for the secular trend. A United Nations study published in 1953 attributed the trend in mortality to four categories of advances: (1) public-health reforms; (2) advances in medical knowledge and practices; (3) improved personal hygiene; and (4) rising income and standards of living. A United Nations study published in 1973 added 'natural factors', such as the decline in the virulence of pathogens, as an additional explanatory category.

A new phase in the effort to explain the secular decline in mortality was ushered in by Thomas McKeown, who, in a series of papers and books published between 1955 and the mid-1980s, challenged the importance of most of the factors that had previously been advanced for the mortality decline. He was particularly sceptical of those aspects of the consensus explanation that focused primarily on changes in medical technology and public-health reforms. In their place he substituted improved nutrition, but he neglected the synergism between infection and nutrition and so failed to distinguish between diet and nutrients available for cellular growth. McKeown did not make his case for nutrition directly but largely through a residual argument after having rejected other principal explanations. The debate over the McKeown thesis continued through the beginning of the 1980s. However, during the 1970s and 1980s it was overtaken by the growing debate over the elimination of mortality crises as the principal reason for the first wave of the mortality decline.

The systematic study of mortality crises and their possible link to famines was initiated by Meuvret in 1946. Such work was carried forward in France and numerous other countries on the basis of local studies that made extensive use of parish records. This line of research was greatly accelerated by Goubert's 1960 study of Beauvais (cf. Goubert 1965). By the early 1970s several scores of such studies had been published covering the period from the seventeenth century through the early nineteenth century in England, France, Germany, Switzerland, Spain, Italy, and the Scandinavian countries (Smith 1977; Flinn 1981). The accumulation of local studies provided the foundation for the view that mortality crises accounted for a large part of total mortality during the early modern era, and that the decline in mortality rates between the mid-eighteenth and mid-nineteenth centuries was explained largely by the elimination of these crises, a view that won widespread if not universal support.

It was only after the publication of death rates based on large representative samples of parishes for England and France that it became possible to assess the national impact of crisis mortality on total national mortality.[2] Analyses of these series confirmed one of the important conclusions derived from the local studies: mortality was far more variable before 1750 than afterward. They also revealed that the elimination of crisis mortality, whether related to famines or not, accounted for only a small fraction of the secular decline in mortality rates. About 90 per cent of the drop was due to the reduction of 'normal' mortality (Wrigley and Schofield 1981; Dupâquier 1989; Weir 1989; Fogel 1993b).

In discussing the factors that had kept past mortality rates high, the authors of the 1973 United Nations study of population noted that 'although chronic food shortage has probably been more deadly to man, the effects of famines, being more spectacular, have received greater attention in the literature' (United Nations 1973: 142). Similar points were made by Lebrun (1971) and Flinn (1974, 1981), but it was not until the publication of the INED data for France (Blayo 1975) and the Wrigley and Schofield (1981) data for England, that the limited influence of famines on mortality became apparent (cf. Dupâquier 1989). In chapter 9 of the Wrigley and Schofield volume, Lee (1981) demonstrated that, although there was a statistically significant lagged relationship between large proportionate deviations in grain prices and similar deviations in mortality, the net effect on mortality after five years was negligible. Similar results were obtained by studies of France (Weir 1982, 1989; Richards 1984; Galloway 1986) and Sweden (Bengtsson and Ohlsson 1984, 1985; Galloway 1987; cf. Eckstein, Schultz, and Wolpin 1984).

By demonstrating that famines and famine mortality are a secondary issue in the escape from the high aggregate mortality of the early modern era, these studies have indirectly pushed to the top of research agendas the issue of chronic malnutrition and its relationship to the secular decline in mortality. It is clear that the new questions cannot be addressed by relating annual deviations of mortality (from the trend line) to annual deviations of supplies of food (from their trend). What is now at issue is how the trend in chronic malnutrition might be related to the trend in mortality and how to identify the factors that determined each of these secular trends.

The current concern with the role of chronic malnutrition in the secular decline of mortality does not represent a return to the belief that the entire secular trend in mortality can be attributed to a single overwhelming factor. Specialists currently working on the problem agree that a range of factors is involved, although they have different views on the relative importance of each of the factors. The unresolved issue, therefore, is how much each of the various factors contributed to the decline. Resolution of the issue is essentially an accounting exercise of a particularly complicated nature that involves measuring not only the direct effect of particular factors but also their indirect effects and their interactions with other factors. An important aspect of the current work on chronic malnutrition is its consistency not only with the synergy between nutrition and infection (Scrimshaw, Taylor, and Gordon 1968; Preston and van de Walle 1978), but also with the search for pathways through which the balance between pathogens and human hosts may have changed, a point that has been emphasized by Fridlizius (1984), Perrenound (1984, 1991),

Alter and Riley (1989), and Schofield and Reher (1991) among others (cf. Fogel 1994).

An unfortunate aspect of some past studies of the contribution of improvements in nutritional status to the secular decline in mortality has been the implicit assumption that diet alone determines nutritional status. However, epidemiologists and nutritionists are careful to distinguish between these terms. To them, nutritional status denotes the balance between the intake of nutrients and the claims against it. It follows that an adequate level of nutrition is not determined solely by diet, which is the level of nutrient intake, but varies with individual circumstances. Whether the diet of a particular individual is nutritionally adequate depends on such matters as his level of physical activity, the climate of the region in which he lives, and the extent of his exposure to various diseases. As Nevin S. Scrimshaw put it, the adequacy of a given level of iron consumption depends critically on whether an individual has hookworm.[3]

Thus, a population's nutritional status may decline at the same time that its consumption of nutrients is rising if the extent of its exposure to infection or the degree of its physical activity is rising even more rapidly. It follows that the assessment of the contribution of nutrition to the decline in mortality requires measures not only of food consumption but also of the balance between food consumption and the claims on that consumption. To avoid confusion I will use the terms 'diet' and 'gross nutrition, to designate nutrient intake only. All other references to nutrition, such as 'nutritional status', 'net nutrition', 'nutrition', 'malnutrition', and 'undernutrition' will designate the balance between nutrient intake and the claims on that intake.

1.1 Energy Cost Accounting and Size Distributions of Calories

In developed countries today, and even more so in the less developed nations of both the past and the present, the basal metabolic rate (BMR) is the principal component of the total energy requirement. The BMR—which varies with age, sex, and body weight—is the amount of energy required to maintain the body while at rest; it is the amount of energy required to maintain body temperature and to sustain the functioning of the heart, liver, brain, and other organs. For adult males aged 20–39 living today in moderate climates, BMR normally ranges between 1,350 and 2,000 kcal per day, depending on height and weight (Quenouille *et al.* 1951: 71–2; Davidson *et al.* 1979: 19–25; FAO/WHO/UNU 1985), and for reasonably well-fed persons normally represents somewhere in the range of 45–65 per cent of total calorie requirements (FAO/WHO/UNU 1985: 71–7). Because the BMR does not allow for the energy required to eat and digest food, nor for essential hygiene, an individual cannot survive on the calories needed for basal metabolism. The energy required for these additional essential activities over a period of twenty-four hours is estimated at 0.27 of BMR or 0.4 of BMR during waking hours. In other words, a survival diet is 1.27 BMR.[4] Such a diet, it should be emphasized, contains no allowance for the energy required to earn a living, prepare food, or any movements beyond those connected with eating and essential hygiene. It is not sufficient to maintain long-term health but represents the short-term maintenance level 'of totally inactive dependent people' (FAO/WHO/UNU 1985: 73).

Energy requirements beyond maintenance depend primarily on how individuals spend their time besides sleeping, eating, and essential hygiene. This residual time will normally be divided between work and such discretionary activities as walking, community activities, games, optional household tasks, and athletics or other forms of exercise. The energy requirements of a large number of specific activities (expressed as a multiple of the BMR requirement per minute of an activity) have been worked out (see Table 9.1 for some examples). In order to standardize for the age and sex distribution of a population, it is convenient to convert the per-capita consumption of calories into consumption per equivalent adult male aged 20–39 (which is referred to as a 'consuming unit'). Energy cost accounting is usually worked forwards, going from a list of activities to an estimate of the average daily caloric requirement, but such accounting can also be worked backwards, going from the average caloric intake to the residual (after deducting the survival level of energy) available for work and discretionary activities.

Size distributions of caloric consumption are one of the most potent instruments used in assessing the plausibility of proffered estimates of average diets. They not only bear on the implications of a given level of caloric consumption for morbidity and mortality rates, but they also indicate whether the calories available for work are consistent with the level of agricultural output and with the distribution of the labour force bet-

TABLE 9.1 *Examples of the energy requirements of common activities expressed as a multiple of the basal metabolic rate (BMR) for males and females*

Activity	Males	Females
Sleeping	1.0	1.0 (i.e. BMR x 1.0)
Standing quietly	1.4	1.5
Strolling	2.5	2.4
Walking at normal pace	3.2	3.4
Walking with 10-kg load	3.5	4.0
Walking uphill at normal pace	5.7	4.6
Sitting and sewing	1.5	1.4
Sitting and sharpening machete	2.2	—
Cooking	1.8	1.8
Tailoring	2.5	
Carpentry	3.5	—
Common labour in building trade	5.2	—
Milking cows by hand	2.9	—
Hoeing	—	5.3–7.5
Collecting and spreading manure	6.4	—
Binding sheaves	5.4–7.5	3.3–5.4
Uprooting sweet potatoes	3.5	3.1
Weeding	2.5–5.0	2.9
Ploughing	4.6–6.8	—
Cleaning house	—	2.2
Child care	—	2.2
Threshing	—	4.2
Cutting grass with machete	—	5.0
Laundry work	—	3.4
Felling trees	7.5	—

Note: Rates in Durnin and Passmore (1967: 31) given in kcal/min were converted into multiples of BMR, using kcal per min of a 65 kg man and a 55 kg woman of average build.

Sources: Durnin and Passmore 1967: 31, 66, 67, 72; FAO/WHO/UNU 1985: 76–78, 186–91.

ween agriculture and non-agriculture (Fogel 1991; Fogel and Floud 1994). Although national food balance sheets, such as those constructed by Toutain (1971) for France over the period 1781–1952, provide mean values of per-capita caloric consumption, they do not produce estimates of the size distribution of calories. In principle it is possible to construct size distributions of calories from household consumption surveys.

Inasmuch as most of these surveys during the nineteenth century were focused on the lower classes, in order to make use of them it is necessary to know from what centiles of either the national-caloric or the national-income distribution the surveyed households were drawn.

Three factors make it possible to estimate the size distributions of calories from the patchy evidence available to historians. First, studies covering a wide range of countries indicate that distributions of calories are well described by the lognormal distribution. Secondly, the variation in the distribution of calories (as measured by the coefficient of variation $s/\overline{X}$ or the Gini (G) ratio) is far more limited than the distribution of income. In contradistinction to income, the bottom tail of the caloric distribution is sharply restricted by the requirement for basal metabolism and the prevailing death rate. The bottom end is also constrained by the requirement that the energy available to the agricultural labour force is sufficient to produce the agricultural output. At the top end it is restricted by the human capacity to use energy and the distribution of body builds. Consequently, the extent of the inequality of caloric distributions is pretty well bounded by: $0.4 \geq (s/\overline{X}) \geq 0.2$ ($0.22 \geq G \geq 0.11$) (Aitchison and Brown 1966; FAO 1977; US National Center for Health Statistics 1977; Lipton 1983). Thirdly, when the mean is known, the coefficient of variation (which together with the mean determines the distribution) can be estimated from either tail of the distribution. Even in places and times where little is known about ordinary people, there is a relative abundance of information about the rich. At the bottom end information on mortality rates and on the energy requirement of the agricultural labour force rather tightly constrain the proportion of the distribution that lies below BMR.

1.2 Estimates and Implications of the Levels and Distributions of Caloric Consumption in Britain and France near the End of the *Ancien Régime*

Owning to the work of Toutain (1971) we have a series of estimates of average caloric consumption for France that extends back to *c*.1785. His estimates, which are derived from national food balance sheets, imply that the average caloric consumption in France on the eve of the French Revolution was about 1,753 kcals per capita or about 2,290 kcals per consuming unit. This is a very low level of energy consumption, ranking

France c.1785 with such impoverished countries today as Pakistan and Rwanda (World Bank 1987).

Work on the English caloric consumption has lagged behind that of France. Oddy (1990) and Shammas (1990), using the budgets of rural households collected by Davies for c.1790 and Eden for c.1794 (Stigler 1954), have estimated that the mean daily caloric consumption was in the neighbourhood of 2,100 kcals per capita. Standardizing for age and gender and allowing for the place of these rural households in the English size distribution of income developed by Lindert and Williamson (1982; cf. Fogel 1993*b*) for the mid-eighteenth century, this figure implies a national average daily caloric consumption of about 2,700 kcals per consuming unit.

TABLE 9.2 *Daily caloric consumption in England and Wales*

Date	kcals per capita	kcals per consuming unit
1700	1,802	2,365
1750	1,911	2,508
1800	2,346	3,079
1850	2,235	2,933

Recent work by agricultural historians summarized by Holderness (1989) and Allen (1994) have yielded estimates of English and Welsh agricultural output at half-century intervals between 1700 and 1850. The procedures used to convert these estimates into a national food balance sheet are described in Fogel and Floud (1994). Since work on the refinement of these procedures is still in progress, the estimates of average daily caloric consumption presented in Table 9.2 are provisional and are only rough indicators of the possible course of consumption. They should be treated as illustrative, suggesting the orders of magnitude that are involved and posing provisional hypotheses aimed at stimulating future research. If we interpolate this series geometrically, the estimate of kcals per consuming unit in 1790 is 2,955, which is about 255 kcals more than the figure obtained from the household studies.

One implication of the French and English estimates of mean daily caloric consumption is that mature adults of the late eighteenth century must have been very small by current standards. Today the typical American male in his early thirties is about 177 cm (69.7 inches) tall and weighs about 78 kg (172 lbs) (US Department of Health and Human

Services 1987). Such a male requires daily about 1,794 kcal for basal metabolism and a total of 2,279 kcal for baseline maintenance. If either the British or the French had been that large during the eighteenth century, virtually all of the energy produced by their food supplies would have been required for maintenance and hardly any would have been available to sustain work. To have the energy necessary to produce the national products of these two countries *c*.1700, the typical adult male must have been quite short and very light.

TABLE 9.3 *Estimated average final heights of men who reached maturity between 1750 and 1875 in six European populations by quarter centuries (cm)*

Date of matu-rity by century and quarter	UK	Norway	Sweden	France	Denmark	Hungary
18 III	165.9	163.9	168.1	—	—	168.7
18 IV	167.9	—	166.7	163.0	165.7	165.8
19 I	168.0	—	166.7	164.3	165.4	163.9
19 II	171.6	—	168.0	165.2	166.8	164.2
19 III	169.3	168.6	169.5	165.6	165.3	—
20 III	175.0	178.3	177.6	172.0	176.0	170.9

Sources: Fogel (1987: table 7) for all columns except France. For France, nineteenth-century data were computed from Meerton (1989) as amended by Weir (1993), with 0.9 cm added to allow for additional growth between age 20 and maturity (Gould 1869: 104–105; cf. Friedman 1982: 510 n. 14). The entry to row 2 is derived from a linear extrapolation of Meerton's data for 1815–36 back to 1788, with 0.9 cm added for additional growth between age 20 and maturity. The entry in row 6 is from Fogel (1987: table 7).

This inference is supported by data on stature and weight which have been collected for European nations. Table 9.3 provides estimates of final heights of adult males who reached maturity between 1750 and 1875. It shows that during the eighteenth and nineteenth centuries Europeans were severely stunted by modern standards (cf. row 6 of Table 9.3). Estimates of weights for European nations before 1860 are much more patchy. Those which are available, mostly inferential, suggest that *c*.1790 the average weight of English males in their thirties was about 61 kg (134 lb), which is about 20 per cent below current levels. The corresponding figure for French males *c*.1790 may have been only about 50 kg (about 110 lb), which is about a third below current standards.

Table 9.4 displays the caloric distribution for England and France implied by the available evidence.[5] It shows the exceedingly low level of work capacity permitted by the food supply in France and England c.1790, even after allowing for the reduced requirements for maintenance because of small stature and body mass (cf. Freudenberger and Cummins 1976). In France the bottom 10 per cent of the labour force lacked the energy for regular work and the next 10 per cent had enough energy for less than three hours of light work daily (0.52 hours of heavy work). Although the English situation was somewhat better, the bottom 3 per cent of the labour force lacked the energy for any work, but the balance of the bottom 20 per cent had enough energy for about six hours of light work (1.09 hours of heavy work) each day.

TABLE 9.4 *A comparison of the probable French and English distributions of the daily consumption of kcals per consuming unit towards the end of the eighteenth century*

Decile	A. France c. 1785 $\overline{X} = 2,290\ (s/\overline{X}) = 0.3$		B. England c. 1790 $\overline{X} = 2,700\ (s/\overline{X}) = 0.3$	
	Daily kcal consumption	Cumulative %	Daily kcal consumption	Cumulative %
Highest	3,672	100	4,329	100
Ninth	2,981	84	3,514	84
Eighth	2,676	71	3,155	71
Seventh	2,457	59	2,897	59
Sixth	2,276	48	2,684	48
Fifth	2,114	38	2,492	38
Fourth	1,958	29	2,309	29
Third	1,798	21	2,120	21
Second	1,614	13	1,903	13
First	1,310	6	1,545	6

Sources and procedures: See Fogel (1993b).

Table 9.4 also points up the problem with the assumption that, for *ancien régime* populations, a caloric intake that averaged 2,000 kcal per capita (2,600 per consuming unit) daily was adequate (Livi-Bacci 1990). That average level of consumption falls between the levels experienced by the French and the English c.1790. In populations experiencing such low levels of average consumption, the bottom 20 per cent subsisted on such poor diets that they were effectively excluded from the labour force, with many of them lacking the energy even for a few hours of strolling. That appears to be the principal factor explaining why beggars constituted as much as a fifth of the populations of *ancien régime* France (Laslett 1965; Goubert 1973; Cipolla 1980). Even the majority of those in the top 50 per cent of the caloric distribution were so stunted (height below US standards) and wasted (weight below US standards) that they were at substantially higher risk of incurring chronic health conditions and of premature mortality (see Section 1.4 below).

1.3. Stature and Body Mass Indexes as Indicators of Secular Trends in Morbidity and Mortality

Extensive clinical and epidemiological studies over the past two decades have shown that height at given ages, weight at given ages, and weight-for-height (a body mass index) are effective predictors of the risk of morbidity and mortality (see the summaries in Osmani 1992; Fogel 1993*b*; and Fogel, Costa, and Kim 1994; cf. Heywood 1983; Waaler 1984; Martorell 1985). Height and body-mass indexes measure different aspects of malnutrition and health. Height is a net rather than a gross measure of nutrition. Moreover, although changes in height during the growing years are sensitive to current levels of nutrition, mean final height reflects the accumulated past nutritional experience of individuals over all of their growing years, including the foetal period. Thus, it follows that, when final heights are used to explain differences in adult mortality rates, they reveal the effect, not of adult levels of nutrition on adult mortality rates, but of nutritional levels during infancy, childhood, and adolescence on adult mortality rates. A weight-for-height index, on the other hand, reflects primarily the current nutritional status. It is also a net measure in the sense that a body-mass index (BMI) reflects the balance between intakes and the claims on those intakes. The most widely used body-mass index is weight measured in kilograms divided by height measured in metres squared (kg/m^2), sometimes called the Quetelet index. Although height is determined by the cumulative nutritional status during an entire developmental age span, the BMI fluctuates with the current balance between nutrient intakes and energy demands. A person whose height is short relative to the modern US or West European standard is referred to as 'stunted'. Those with low BMI's are referred to as 'wasted'.

The predictive power of height and body mass-indexes with respect to morbidity and mortality are indicated by Figs. 9.1 and 9.2. Fig. 9.1*a* reproduces a diagram by Waaler (1984). It shows that short Norwegian men aged 40–59 at risk between 1963 and 1979 were much more likely to die than tall men. Indeed, the risk of mortality for men with heights of

165 cm (65.0 inches) was on average 71 per cent greater than that of men who measured 182.5 cm (71.9 inches). Fig. 9.1*b* shows that height is also an important predictor of the relative likelihood that men aged 23–49 would be rejected from the Union Army during 1861–5 because of chronic diseases. Despite significant differences in mean heights, ethnicities, environmental circumstances, the array and severity of diseases, and time, the functional relationship between height and relative risk is strikingly similar in the two diagrams.

Waaler (1984) has also studied the relationship in Norway between BMI and the risk of death in a sample of 1.8 million individuals. Curves summarizing his findings are shown in Fig. 9.2 for both men and women. Although the observed values of the BMI (kg/m^2) ranged between 17 and 39, over 90 per cent of the males had BMIs within the range 21–9. Within the range 23–7, the curve is relatively flat, with the relative risk of mortality hovering close to 1.0. However, at BMIs of less than 21 and over 29, the risk of death rises quite sharply as the BMI moves away from its mean value. It will be noticed that the BMI curves are much more symmetrical than the height curves in Fig. 9.1, which indicates that high BMIs are as risky as low ones.

Adult height and the BMI measure different aspects of nutritional status. Not only is stunting due to malnutrition during developmental ages, but it appears that most stunting occurs under age 3, after which even badly stunted children generally move along a given height centile, that is, develop without incurring further height deficits (Billewicz and MacGregor 1982; Tanner 1982; Horton 1984; Martorell 1985). Secondly, no matter how badly stunted an adult might be, it is still possible to have an optimum (or good) weight for that height. Thus, for example, a Norwegian male stunted by two inches during his developmental ages could still have had a normal risk if his BMI was about 26.

The fact that even badly stunted populations may have quite normal BMIs reflects the capacity of human beings to adapt their behaviour to the limitations of their food supply. Adaptation takes place in three dimensions. Small people have lower basal metabolism, because less energy is needed to maintain body temperature and sustain the function of vital organs. Small people need less food and hence require less energy to consume their food and for vital hygiene. The third aspect of adaptation comes in the curtailment of work and discretionary activity. If a small (56 kg) man confined himself to a few hours of light work each day,

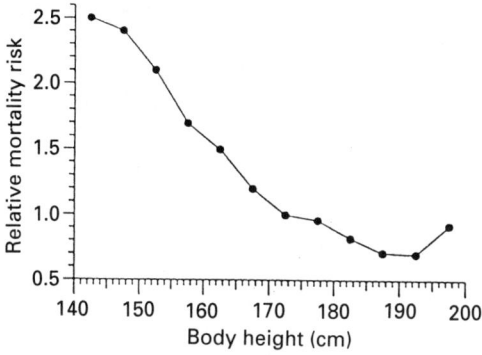

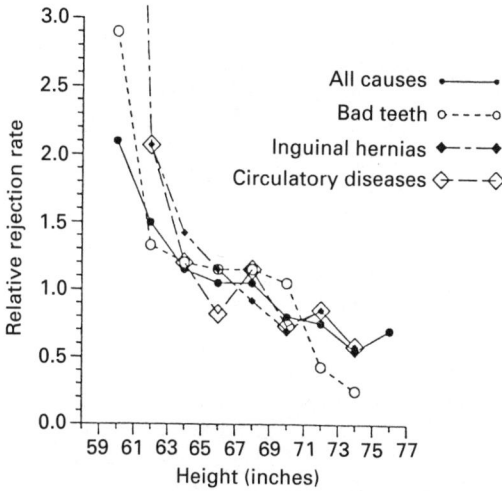

FIG. 9.1 Comparison of the relationship between body height and relative risk in two populations
Source: Waaler (1984); Fogel *et al.* (1986).

246

Robert Fogel

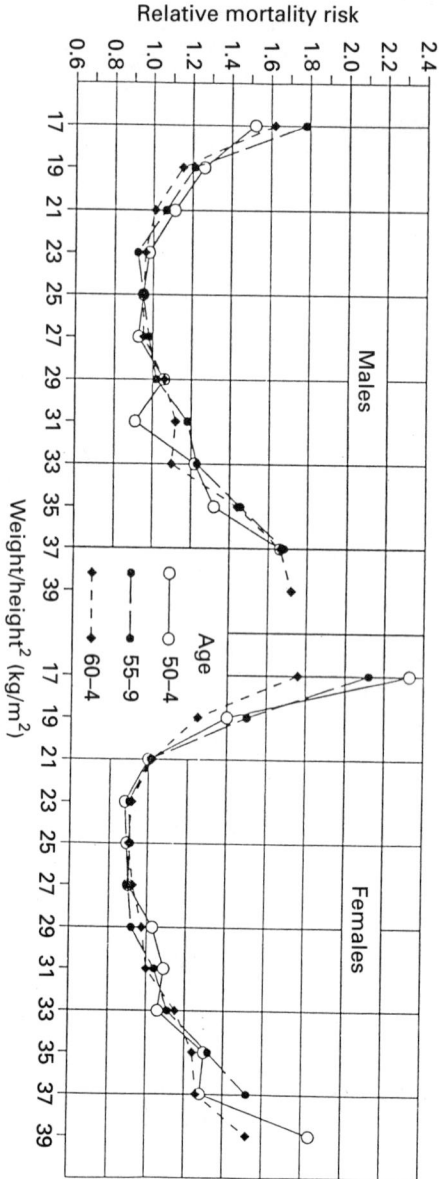

FIG. 9.2 Relationship between BMI and prospective risk among Norwegian adults aged 50–64 at risk, 1963–1979
Source: Waaler (1984).

he could remain in energy balance and maintain his BMI at a satisfactory level with as little as 2,000 or 2,100 kcals. However, a larger man (79 kg) engaged in heavy work for eight hours per day would require about 4,030 kcals to maintain his energy balance at a BMI of 24 (FAO/WHO/UNU 1985).

Although Figs. 9.1 and 9.2 are revealing, neither one singly, nor both together, are sufficient to shed light on the debate over whether moderate stunting impairs health when weight-for-height is adequate, since Fig. 9.1 is not controlled for weight and Fig. 9.2 is only partially controlled for height (Fogel 1987; Fogel and Floud 1994). To get at the 'small-but-healthy' (Seckler 1982) issue one needs an iso-mortality surface that relates the risk of death to both height and weight simultaneously. Such a surface, presented in Fig. 9.3, was fitted to Waaler's data by a procedure described elsewhere (Fogel 1993*b*; Kim 1993). Transecting the iso-mortality map are lines which give the locus of BMI between 16 and 34, and a curve giving the weights that minimize risk at each height.

Fig. 9.3 shows that, even when body weight is maintained at what Fig. 9.2 indicates is an 'ideal' level (BMI = 25), short men are at substantially greater risk of death than tall men. Thus, an adult male with a BMI of 25 who is 164 cm tall is at about 55 per cent greater risk of death than a male at 183 cm who also has a BMI of 25. Fig. 9.3 also shows that the 'ideal' BMI (the BMI that minimizes the risk of death) varies with height. A BMI of 25 is 'ideal' for men in the neighborhood of 176 cm, but for tall men (greater than 183 cm) the ideal BMI is between 22 and 24, while for short men (under 168 cm) the 'ideal' BMI is about 26.

Before using Waaler surfaces to evaluate the relationship between chronic malnutrition and the secular decline in mortality rates since 1700, several issues in the interpretation of that figure need to be addressed. First, it is important to understand the physiological foundation for the predictive capacity of Waaler surfaces and curves. Although research in this area is still developing rapidly and some of the new findings are yet to be confirmed, variations in height and weight appear to be associated with variations in the chemical composition of the tissues that make up the organs of the body, in the quality of electrical transmission across membranes, and in the functioning of the endocrine system and other vital systems. Stunting and other physiological impairments that take place *in utero* or in early childhood are sometimes promptly visible, as in the cases of Foetal Alcohol Syndrome and severe protein calorie mal-

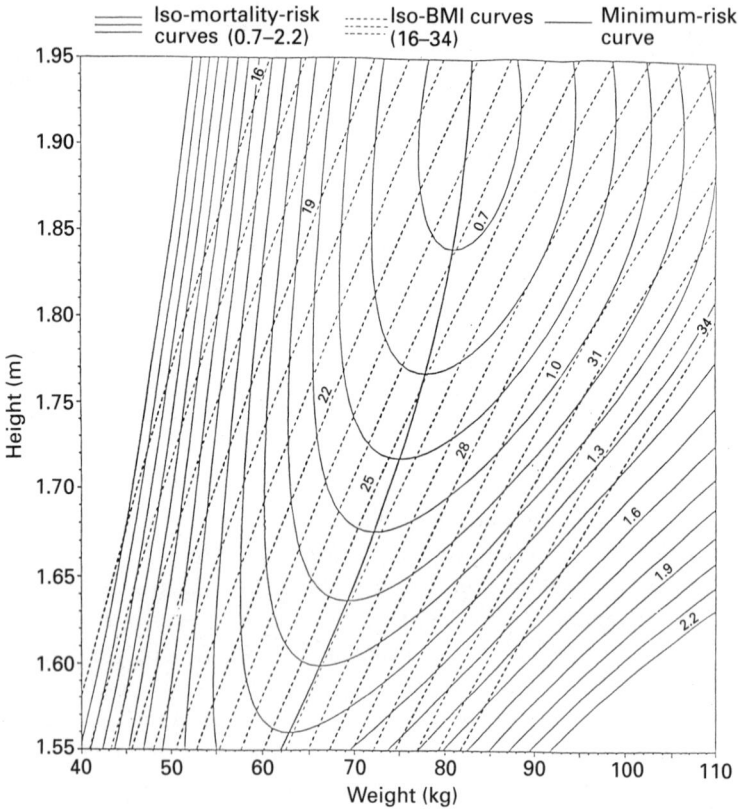

FIG. 9.3 Iso-mortality curves of relative risk for height and weight among Norwegian males aged 50–64, 1963-1979

Note: All risks are measured relative to the average risk of mortality (calculated over all heights and weights) among Norwegian males aged 50–64.

nutrition, both of which can lead to a permanent impairment of central nervous system function. Other physiological impairment caused by malnutrition *in utero* or early childhood may not show up until middle and late ages, when they increase the risk of incurring such conditions as coronary heart disease, hypertension, stroke, diabetes, and autoimmune thyroiditis. Persistent malnutrition at any point in the life cycle can produce severe physiological dysfunctions. In the case of the respiratory system, for example, there is not only decreased muscle mass and strength but also impaired ventilatory drive, biochemical changes in the connective system, and electrolyte abnormalities (Robbins, Cotran, and Kumar 1984; McMahon and Bistrian 1990; Martorell, Rivera, and Kaplowitz 1990; Barker 1993; cf. Fogel 1994).

Secondly, since an individual's height cannot be varied by changes in nutrition after maturity, adults can move to a more desirable BMI only by changing weight. I, therefore, interpret the *X* axis as a measure of the effect of the current nutritional status of mature males on adult mortality rates. Moreover, since most stunting takes place below age 3 (Tanner 1982; Horton 1984; Martorell 1985; Steckel 1986), I interpret the *Y* axis as a measure of the effect of nutritional deprivation *in utero* or early in childhood on the risk of mortality at middle and late ages (cf. Tanner 1982; Steckel 1987; Fogel, Galantine, and Manning 1992: chs. 42, 47).

Thirdly, in applying Fig. 9.3 to the evaluation of secular trends in nutrition and mortality, I assume that for Europeans environmental factors have been decisive in explaining the secular increase in heights, not only for population means, but also for individuals in particular families. The reasonableness of this assumption becomes evident when one considers the issue of shortness. If shortness is defined as a given number of standard deviations below a changing mean (that is, short is 2 standard deviations below the mean, whether the mean is 164 cm or 183 cm), then genetic and environmental factors may be difficult to disentangle. If, however, shortness is defined in absolute terms, say as applying to all males with heights below 168 cm, then it is quite clear that most shortness in Europe and America during the eighteenth and much of the nineteenth centuries was determined by environmental rather than genetic factors.

The point at issue can be clarified by considering the experience of the Netherlands. Shortness has virtually disappeared from that country during the past century and a half. Today, less than 2 per cent of young adult males are below 168 cm, but in *c*.1855 about two-thirds were below that height. Since there has been little change in the gene pool of the Dutch during the period, it must have been changes in environmental circumstances, nutrition, and health that eliminated about 95 per cent of all short males from the Dutch population (Van Wieringen 1986; Fogel 1987). Given current growth rates in the mean final height of the Netherlands, the remaining men shorter than 168 cm may yet be virtually eliminated from the Dutch population.

The Dutch case illustrates the general secular pattern of physical growth in the nations of Western Europe. The secular increase in mean final heights, which ranged between 10 and 20 cm (between 4 and 8 inches) over the past 200 years, cannot be attributed to natural selection or genetic drift, since these processes require much longer time spans. Nor can it be attributed to heterosis (hybrid vigour), because the populations in question have remained relatively homogeneous and because the effects of heterosis in human populations have been shown both empirically and theoretically to be quite small (Damon 1965; Cavalli-Sforza and Bodmer 1971; Van Wieringen 1978; Fogel *et al.* 1983; Martorell 1985; Mueller 1986). Only the top 6 per cent of the Dutch height distribution of *c*.1855 overlaps with the bottom 6 per cent of the current distribution of final heights. Since the Dutch mean is still increasing, and we do not yet know the maximum mean genetically obtainable (often referred to as the genetic potential), it may well be that even the 6 per cent overlap between the distribution of final heights in the *c*.1855 generation and that of the latest generation will be cut in the next few decades, perhaps by as much as half.

Fourthly, even if the Norwegian iso-mortality surface is applicable to European populations generally, the surface may not have been stable over time. Since height-specific and weight-specific mortality rates are measured relative to the average death rate for the population as a whole, short-term shifts in average death rates by themselves will not necessarily shift the surface. However, fundamental shifts in environment, including changes in medical technology, may change the risk surface.[6]

1.4 How Variations in Body Size Brought the Population and the Food
Supply into Balance and Determined the Level of Mortality

The implication of energy cost accounting permitted by the recent work
of agricultural historians is that to have the energy necessary to produce
the national products of England and France *c*.1700, the typical adult
male must have been quite short and very light, weighing perhaps 25–35
per cent less than his American counterpart today. How Europeans of the
past may have adapted their body size to accommodate their food supply
is illustrated by Table 9.5, which compares the estimated average daily
consumption of calories in England and Wales in 1700 and 1800 by two
economic sectors: agriculture and everything else. Within each sector the
estimated amount of dietary energy required for work is also shown. Row
3 presents a measure of the efficiency of the agricultural sector in the
production of dietary energy. That measure is the number of calories of
food output per calorie of work input.

Column 1 of the table presents the situation in 1800, when kcals available for consumption were relatively high by prevailing European
standards (about 3,079 per consuming unit daily), when adult male stature made the British the tallest national population in Europe (168 cm or
66.1 inches at maturity) and relatively heavy by the prevailing European
standards, averaging about 63.5 kg (about 140 pounds) at prime working
ages, which implies a BMI of about 22.5. Food was abundant because in

TABLE 9.5 *A comparison of the average daily uses of dietary energy in England and Wales
in 1700 and 1800*

	1800	1700	1700 counterfactual
Total daily dietary energy consumed (production plus net imports) (kcal m.)	21,480	10,498	7,825
Energy used to produce agricultural output (kcal m.)	959	515	378
Energy productivity in agriculture (the output/input ratio of dietary energy)	21.4	20.7	20.7
Energy consumed in the agricultural sector (kcal m.)	8,722	6,076	5,772
Energy consumed outside of the agricultural sector (kcal m.)	12,758	4,422	2,053
Energy used to produce nonagricultural output (kcal m.)	1,397	444	0

Note: See Fogel and Floud 1994 for a discussion of the sources and procedures involved in the construction of this table.

addition to a substantial domestic production Britain imported about 5 per cent of its dietary consumption. However, as column 1 indicates, British agriculture was quite productive. English and Welsh farmers pro- duced 21.4 calories of food output (net of seeds, feed, inventory losses, and so on) for each calorie of their work input. About 42 per cent of this bountiful output was consumed by the families of the agriculturalists $(8,722 / 20,559 \approx 0.42)$.

The balance of their dietary output, together with some food imports, was consumed by the non-agricultural sector, which constituted about 64 per cent of the English population in 1801 (Wrigley 1987: 170). Al- though food consumption per capita was about 8 per cent lower in this sector than in agriculture, most of the difference was explained by the greater caloric demands of agricultural labour.[7] Food was so abundant that even the English paupers and vagrants, who accounted for about 12 per cent of the population c.1800 (Lindert and Williamson 1982), had about three times as much energy for begging and other activities beyond maintenance as did their French counterparts (Fogel and Floud 1994).

The food situation was much tighter in 1700, when only about 2,365 kcals were available daily per consuming unit. The adjustment to the lower food supply was made in three ways. First, compared to 1800 the share of dietary energy made available to the non-agricultural sector in 1700 was lower by nearly two-thirds, a level that was maintained partly by constraining the share of the labour force engaged outside agriculture. Secondly, the amount of energy available for work per equivalent adult worker was kept low both inside and outside agriculture, although the reduction was somewhat greater outside agriculture. Thirdly, the energy required for basal metabolism and maintenance was low because body size was small. Compared with 1800, adult heights of males of 1700 were down by 5 cm, their BMI was 20 instead of 22.5, and their weights were down by about 10 kg. As a result of such constriction of the average body size of the population, the number of calories required for maintenance was reduced by 170 kcal per consuming unit daily.

The last figure may seem rather small, accounting for just a quarter $(170 / 714 \approx 0.24)$ of the total shortfall in daily caloric consumption, so small that it may seem to undermine the proposition that variations in body size were a principal means of adjusting the population to varia- tions in the food supply. Nevertheless, further consideration of Table 9.5 will sustain the proposition. The condition for a population to be in equilibrium with its food supply at a given level of consumption is that the labour input (measured in calories of work) is large enough to produce the requisite amount of food (also measured in calories). Moreover, a given reduction in calories required for maintenance will have a multiplied effect on the number of calories made available for work. The multiplier is the inverse of the labour-force participation rate. Since only about 35 per cent of equivalent adults were in the labour force, the daily gain in kcals for work was, not 170 per equivalent adult worker, but 485 kcals per equivalent adult worker.

The importance of the last point is indicated by considering columns 2 and 3 of Table 9.5. Column 2 shows that the daily total of dietary energy used for work in 1700 was 959 million kcals, with 515 million expended in agriculture and the balance in non-agriculture. Column 3 indicates what would have happened if all the other adjustments had been made but body size remained at the 1800 level, so that maintenance requirements were unchanged. The first thing to note is that energy available for food production would have declined by 27 per cent. Assuming the same input–output ratio, the national supply of dietary energy would have declined to 7,825 million kcal, of which nearly three-quarters would have been consumed within the agricultural sector. The residual available for non-agriculture would hardly have covered the maintenance requirements of that sector, leaving zero energy for work in non-agriculture. In this example, the failure to have constrained body size would have reduced the energy for work by about 61 per cent (1– (378 / 959)≈0.61).[8]

Varying body size was a universal way that the chronically malnourished populations of Europe responded to food constraints. Such variation in height is evident in Table 9.3 and has been discussed in much more detail for England by Floud, Wachter, and Gregory (1990) and for Hungary by Komlos (1990). Some may want to debate whether the size mechanism was more important than variations in fertility in equating population and the food supply. That interesting question should be pursued, but here I want to focus on the implication of the size mechanism for the explanation of the secular decline in mortality.[9]

Fig. 9.4 superimposes rough estimates of heights and weights in France at four dates on a Waaler surface. In 1705 the food supply in France was even lower than in Britain so that it was necessary to keep body mass even lower than in Britain. In *c.* 1705 the French are estimated

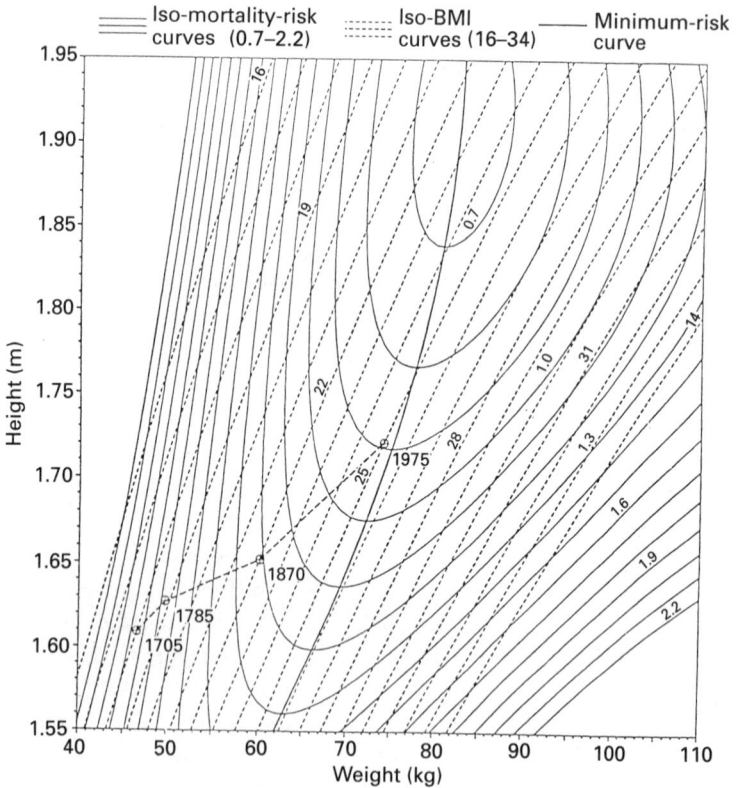

FIG. 9.4 Iso-mortality curves of relative risk for height and weight among Norwegian males aged 50–64, 1963–1979, with a plot of the estimated French height and weight at four dates ·

to have achieved equilibrium with their food supply at a height of about 161 cm and a BMI of about 18. Over the next 270 years the food supply expanded with sufficient rapidity so that both the height and the weight of adult males increased. However, weight appears to have increased more rapidly than height during the first 165 years. Fig. 9.4 indicates that it was factors associated with the gain in BMI that accounted for most of the reduction in the risk of mortality before 1870. After 1870 factors associated with the gain in height explain most of the additional mortality decline predicted by Fig. 9.4.

There is another implication of Fig. 9.4 that is worth making explicit. If the relative risks of mortality in Fig. 9.4 are standardized on the French crude death rate of *c*.1785, one obtains the time series of crude death rates (per thousand) shown in Table 9.6:

TABLE 9.6 *Relative risks of mortality*

| Date (approx.) | Death rates per thousand | |
	Estimated from Fig. 9.4	From registrations or samples
1705	40	—
1785	36	36
1870	26	25
1975	19	11

It thus appears that while factors associated with height and BMI jointly explain about 90 per cent of the decline in French mortality rates over the period between *c*.1785 and *c*.1870, they only explain about 50 per cent of the decline in mortality rates during the past century. Increases in body size continued to be a major indicator of improved life expectation among persons of relatively good nutritional status, but during the last century factors other than those which act through height and BMI became increasingly important.

The analysis in this section points to the misleading nature of the concept of subsistence as Malthus originally used it and as it is still widely used today. Subsistence is not located at the edge of a nutritional cliff, beyond which lies demographic disaster. The evidence outlined in the chapter implies that, rather than one level of subsistence, there are numerous levels at which a population and a food supply can be in equilibrium, in the sense that they can be indefinitely sustained. However, some levels will have smaller people and higher 'normal' (non-crisis) mortality than others. Moreover, with a given population and technology one can alter body size and mortality by changing the allocation of labour between agriculture and other sectors. Thus, the larger the share of the labour force that is in agriculture, all other things being equal, the larger the share of caloric production that can be devoted to baseline maintenance.[10]

Although there was a wide range of levels at which the population and the food supply could have been in equilibrium, not all of these equilibria were equally desirable. Some equilibria left those in the bottom portion of the caloric distribution with so little energy that as much as a quarter of the potential labour force was effectively excluded from production. Such equilibria also made the population highly vulnerable to periodic breakdowns in the food distribution system (famines), although chronic malnutrition was responsible for many more deaths than the famines that called attention to the plight of the lower classes. Some equilibria required 80 per cent or more of the labour force to work in the agricultural sector. However, a fairly high degree of diversification into trade and industry was possible even at fairly low levels of average caloric production. Although England in *c.*1700 and France in *c.*1785 had similar levels of caloric consumption per consuming unit, England was able to support about 45 per cent of its labour force in non-agricultural pursuits, while France supported about 40 per cent of its labour force in such pursuits. The critical difference was in the relatively high output-input ratio of dietary energy in England's agricultural sector (O'Brien and Keyder 1978; Chartres 1985; Wrigley 1987; Holderness 1989; Grantham 1990; Fogel 1993*b*; Allen 1994).

2. MALTHUSIAN ISSUES IN THE LATE TWENTIETH CENTURY: THE RELEVANCE OF THE SIZE MECHANISM FOR THE EVALUATION OF THE CHRONIC MALNUTRITION, WORK CAPACITY, MORBIDITY, AND MORTALITY IN THIRD WORLD COUNTRIES

Malthusian theory still influences current policy debates. There are two sets of issues that could be considered within the analytical framework that has emerged from studies of the secular decline in mortality. One set of issues concerns policies aimed at improving conditions of life in the

Third World. The other set stems from the concern that the ecosystem is threatened by the doubling of the world's population between 1950 and 1990 and the expectation that the population will double again during the next half century (*The Economist* 1990; Livi-Bacci 1992). Since threats to the ecosystem are treated at length elsewhere in this volume, the discussion here is limited to the first set of policy issues.

There are significant parallels between the level of malnutrition in Western Europe *c.*1790 and the poor countries of the Third World during recent decades. The average daily caloric intake of low-income economies in 1990 was about 1,975 kcals per capita (World Bank 1992: 272), which falls between the levels of caloric consumption in France and England two centuries ago. The mean stature of males at maturity in these Third World countries was about 163 cm and the mean BMI was in the range of 18–22 (Eveleth and Tanner 1976), figures which again match those estimated for Western Europe near the beginning of the nineteenth century. Neither Malthus nor anyone else complained that the average body builds of their day were excessively low, although they did express alarm about the urban poor who were several centimetres shorter and who probably were about 10 kg lighter at maturity (Floud and Wachter 1982). Moreover, standards for the adequacy of the diet were set with reference to what was normally consumed by persons near the middle of the caloric distribution (Fogel 1989). However, given the work requirement and the exposure to disease, such diets led to equilibrium stature and BMI that substantially increased morbidity and death rates.

Poor body builds increased vulnerability to diseases, not just infectious diseases, but chronic diseases as well. This point is implicit in Fig. 9.1*b*, which shows that chronic conditions were much more frequent among short young men in the 1860s than among tall men. Fig. 9.5 shows that the same relationship between ill health and stature exists among the males covered by the US National Health Interview Surveys (NHIS) for 1985–8. Stunting during developmental ages had a long reach and increased the likelihood that people would suffer from chronic diseases at middle and at late ages.

American males born during the second quarter of the nineteenth century were not only stunted by today's standards, but were also wasted. Their BMIs at adult ages were about 15 per cent lower than current US levels (Fogel, Costa, and Kim 1994). The implication of the combined stunting and wasting is brought out by Fig. 9.6, which presents a Waaler

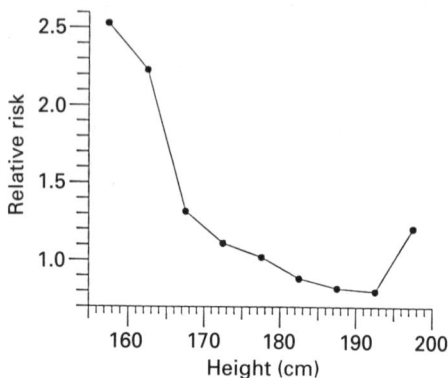

FIG. 9.5 The relationship between height and relative risk of ill health in NHIS veterans aged 40–59, 1985–1988
Source: Fogel, Costa, and Kim (1994).

surface for morbidity estimated by Kim (1993) from NHIS data for 1985–8.

The Waaler surface for risk from chronic conditions in Fig. 9.6 is *similar* to the Norwegian surface for mortality (see Fig. 9.4). The word 'similar' is emphasized because there are some differences that need to be noted. The iso-morbidity curves in ill health rise more steeply than the iso-mortality curves as one moves away in either direction from the optimal weight curve. Furthermore, the optimal weight curve in Fig. 9.6 usually lies about one iso-BMI curve to the right of the optimal weight curve computed from the Norwegian mortality data. Thus, both the Norwegian mortality data and the US health data indicate that for men in the neighbourhood of 1.60–1.65 meters the optimal BMI is in the range of 25 to 27. This is above current levels recommended by FAO/WHO/UNU (1985), falling into the lower ranges of overweight in that standard.

Fig. 9.6 also presents the co-ordinates in height and BMI of Union Army veterans who were 65 or over in 1910 and of veterans (mainly of the Second World War) who were the same ages during 1985–8. These co-ordinates predict a decline of about 35 per cent in the prevalence of chronic disease among the two cohorts. About 61 per cent of the predic-

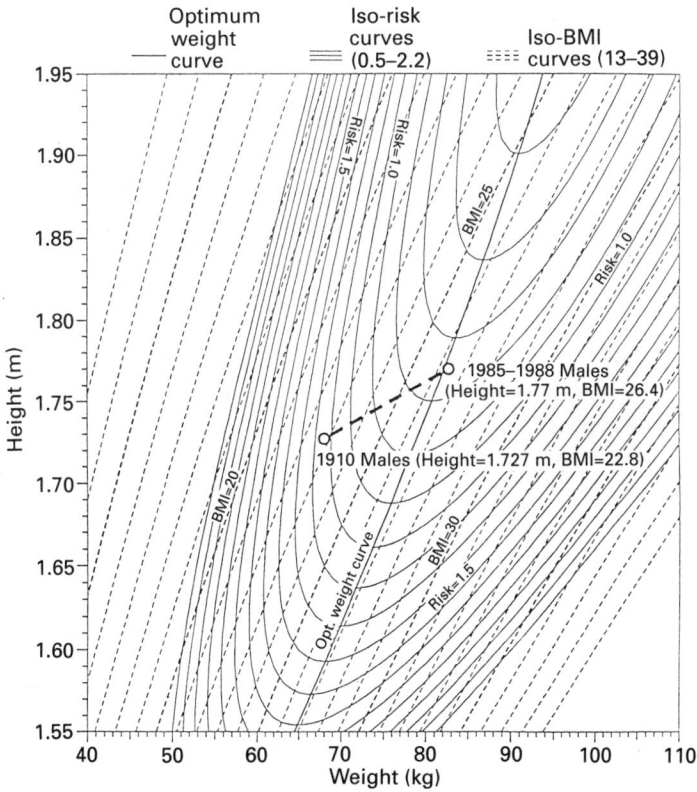

FIG. 9.6 Health improvement predicted by NHIS 1985–1988 health surface
Source: Kim (1993)

ted decline in ill health is due to factors associated with the increase in BMI, and the balance is due to factors associated with increased stature. The decline in the prevalence of chronic diseases indicated by Fig. 9.6 is quite close to what actually occurred. Table 9.7 compares the prevalence of chronic diseases among Union Army men aged 65 and over in 1910 with two surveys of veterans of the same ages in the 1980s. That table indicates that heart disease was 2.9 times as prevalent, musculoskeletal and respiratory diseases were 1.6 times as prevalent, and digestive diseases were 4.7 times as prevalent among veterans aged 65 or over in 1910 as in 1985–8. During the 7.6 decades separating the two groups, the prevalence of heart disease among the elderly declined at a rate of 12.8 per cent per decade, while musculo-skeletal and respiratory diseases each declined at a rate of 5.9 per cent per decade.

TABLE 9.7 *Comparison of the prevalence of chronic conditions among Union Army veterans in 1910, veterans in 1983 (reporting whether they ever had specific chronic conditions), and veterans in NHIS 1985–1988 (reporting whether they had specific chronic conditions during the preceding 12 months), aged 65 and above (%)*

Disorder	1910 Union Army veterans	1983 veterans	Age-adjusted 1983 veterans	NHIS 1985–8 veterans
Musculo-skeletal	67.7	47.9	47.2	42.5
Digestive	84.0	49.0	48.9	18.0
Hernia	34.5	27.3	26.7	6.6
Diarrhoea	31.9	3.7	4.2	1.4
Genito-urinary	27.3	36.3	32.3	8.9
Central nervous, endocrine, metabolic, or blood	24.2	29.9	29.1	12.6
Circulatory[a]	90.1	42.9	39.9	40.0
Heart	76.0	38.5	39.9	26.6
Varicose veins	38.5	8.7	8.3	5.3
Haemorrhoids[b]	44.4			7.2
Respiratory	42.2	29.8	28.1	26.5

Note: Prevailing rates for Union Army veterans are based on examinations by physicians. Those for the 1980s are based on self-reporting. Comparison of the NHIS rates with those obtained from physicians' examinations in NHANES II indicates that use of self-reported health conditions does not introduce a significant bias into the comparison. See the source for a more detailed discussion of possible biases and their magnitudes.

[a] Among veterans in 1983, the prevalence of all types of circulatory diseases will be underestimated because of under-reporting of haemorrhoids.

[b] The variable indicating if the 1983 veteran ever had haemorrhoids is unreliable.

Source: Fogel, Costa , and Kim (1994).

The cohorts born between 1822 and 1845 not only suffered high rates of chronic conditions in old age, they also suffered high rates of chronic diseases at young adult ages (Fogel, Costa, and Kim 1994). Those who survived the deadly infectious diseases of childhood and adolescence were not freer of degenerative diseases, as some have suggested, but more afflicted. At ages 35–9 hernia rates, for example, were more than three times as prevalent in the 1860s as in the 1980s. Of special note is the much higher incidence of clubfoot in the 1860s, a birth anomaly which suggests that the uterus was far less safe for those awaiting birth than it is today. Nor is the idea that violence has increased borne out by the new data. As is indicated by the greater prevalence of deformities of the hand, and other injuries, the mid-nineteenth century was filled with violence that is less often encountered today, particularly in the occupational realm.

Those who survived the killer contagious diseases of early and middle ages were more afflicted by degenerative chronic conditions at old ages in the 1910s than in the 1980s. Nearly 74 per cent of the elderly Union Army veterans suffered from three or more disabling chronic conditions, which is much higher than the rate among elderly veterans in 1983 (Fogel, Costa, and Kim 1994). It may be true that there were fewer genetically frail persons among those who survived to age 65 in 1910 than there are today. If so, that genetic advantage was apparently offset by a life-time of socio-economic and biomedical stress that left health in old age badly impaired and that sharply curtailed the life expectations of the elderly. During the 1910s the elderly died not from the infectious diseases that killed the great majority of their cohorts at relatively young ages but primarily from degenerative diseases which, at the two-digit level, are similar to the distribution of causes of death during the 1980s, except that deaths from neoplasms were lower and deaths from tuberculosis were higher than in the 1980s (cf. Preston 1976).

The provisional findings thus suggest that chronic conditions were far more prevalent throughout the life cycle for those born between 1820 and 1850 than is suggested by the theory of the epidemiological transition. Reliance on causes-of-death information to characterize the epidemiology of the past has led to a significant misrepresentation of the distribution of health conditions among the living. It has also promoted the view that the epidemiology of chronic diseases is more separate from that of infectious diseases than may be the case.

Malthus's legacy to current discussions of population policy and mal-nutrition thus has two aspects, one of which is widely recognized. The other aspect, however, is more covert, although its influence on current policy discussions is powerful. The continued pressure of population against food resources is the acknowledged part of the Malthusian leg-acy. The recently published *Second Report on the World Nutrition Situation* (United Nations 1992), for example, concludes that one-third of the children of the Third World (184,000,000 children) are malnouris-hed. The same report concludes that 40 per cent of women of child-bearing age in South and South-East Asia are wasted (BMI below 18.5) and that between 15 and 20 per cent of these women are stunted (height below 1.45 m).

The other aspect of the Malthusian legacy is that malnutrition mani-fests itself primarily among the exceptional: in the starvation and excess mortality brought on by famines and prevalent among the ultra-poor of his day who lived in misery and vice. Malthus thought that persons near the middle of the social order, the sturdy agricultural labourer or the town artisan, were generally well fed, healthy, and lived normal lives. We now know, however, that famines accounted for less than 4 per cent of the premature mortality of Malthus's age, and that the excess mortality of the ultra-poor (the bottom fifth of society) accounted for another sixth of premature mortality. About two-thirds of all premature mortality in Malt-hus's time came from the part of society that Malthus viewed as productive and healthy. Yet, by current standards, even persons in the top half of the income distribution in England during the eighteenth century were stunted and wasted, suffered far more extensively from chronic diseases at young adult and middle ages than is true today, and died thirty years sooner than today.[11]

This aspect of the Malthusian legacy is embodied in such theses as 'small but healthy', which holds that stunted or wasted individuals may not be more vulnerable to ill health and mortality than those who con-form to the standards of the National Centers for Health Statistics (NCHS). The thesis also calls into doubt the proposition that stunting, in the absence of wasting, affects work capacity at adult ages. This Malthu-sian perspective has arisen partly because of the paucity of life-cycle data sets, especially in Third World countries, which caused investigators to focus only on the early years of the life-span, searching for interaction between natal and infant measures of size and measures of health and work capacity later in childhood. Such studies generally picked up the effects of only exceedingly severe stunting and wasting (more than 2 standard deviations below average), missing the impact of more moderate size effects, many of which do not show up until later in life (Seckler 1980; Sukhatme 1981; Lipton 1983; cf. West *et al.* 1990).

However, the information reported in this chapter indicates that stunting (most of which occurs before age 3) has a long reach, predicting chronic disease rates at young adult and later ages. Low BMI also predicts chronic diseases, both contemporaneously and in later ages. In both the Norwegian and the Union Army data, low BMI at middle ages predicts the odds of dying over long periods after measurement (Costa 1993). Other recently reported life-cycle interactions related to anthropometric measures include the finding that high blood pressure among men and women aged 46 to 54 is negatively related to their birth weight and positively related to placental weight. It has also been reported that mortality from ischaemic heart disease among men in late middle ages was positively related both to their birth weight and to their weight at age 1 (Barker 1991; cf. Elo and Preston 1992; Mackenbach 1992).

The higher prevalence of disabling chronic diseases (such as hernias, chronic diarrhoea, arthritis, heart disease, and respiratory diseases) among stunted and underweight adults in the Third World has escaped attention because the relevant information on such conditions is not generally collected. But the existence of such a relationship in rich countries, both now and when they were much poorer than they are today, suggests that such conditions also exist in Third World countries.

Chronic diseases are not the only way that the existence of chronic malnutrition reduces the productivity of the labour force. When the mean amount of calories is as low as it is in the poor nations of the world, labour-force participation rates and measures of labour productivity are bound to be low, especially when the hours of labour are not adjusted for the intensity of labour (see Fogel 1991). The point at issue here can be illustrated by considering the contribution of improved nutrition to the growth of per-capita income in England between 1790 and 1980. Consideration of the issue starts with the first law of thermodynamics, which applies as strictly to the human engine as to mechanical engines. Since, moreover, the overwhelming share of calories consumed among malnourished populations is required for BMR and essential maintenance, it is quite clear that in energy-poor populations, such as those of Europe during the second half of the eighteenth century, the typical individual in the labour force had relatively small amounts of energy available for work. This observation does not preclude the possibility that malnourished French peasants worked hard for relatively long hours at certain times of the year, as at harvest time. Such work could have been sustained either by consuming more calories than normal during such periods, or by drawing on body mass to provide the needed energy. That level of work, however, could not have been sustained over the entire year. On average, the median individual in the French caloric distribution of 18–IV had only enough energy, over and above maintenance, to sustain regularly about 1.9 hours of heavy work or about 3.4 hours of moderate work per day (Fogel and Floud 1994).

It is quite clear, then, that the increase in the amount of calories available for work over the past 200 years must have made a non-trivial contribution to the growth rate of the per-capita income of countries such as France and Great Britain. That contribution had two effects. First, it increased the labour-force participation rate by bringing into the labour force the bottom 20 per cent of the consuming units, who, even assuming highly stunted individuals and low BMIs, had only enough energy above maintenance for a few hours of strolling each day—about the amount needed for a career in begging—but less on average than that needed for just one hour of heavy manual labour.[12] Consequently, merely the elimination of the large class of paupers and beggars, which was accomplished in England mainly during the last half of the nineteenth century (Lindert and Williamson 1982, 1983; Himmelfarb 1983; Williamson 1985), contributed significantly to the growth of national product. The increase in the labour-force participation rate made possible by raising the nutrition of the bottom fifth of consuming units above the threshold required for work, by itself, contributed 0.11 percentage points to the annual British growth rate between 1780 and 1980 ($1.25^{0.005} - 1 = 0.0011$).

In addition to raising the labour force participation rate, the increased supply of calories raised the average consumption of calories by those in the labour force from 2,944 kcal per consuming unit in c.1790 to 3,701 kcal per consuming unit in 1980. Of these amounts, 1,009 kcal were available for work in c.1790 and 1,569 in 1980, so that calories available for work increased by about 56 per cent during the past two centuries. We do not know exactly how this supply of energy was divided between discretionary activities and work c.1790, but we do know that the pre–

industrial and early industrial routine had numerous holidays, absentee days, and short days (Thompson 1967; Landes 1969). If it is assumed that the proportion of the available energy devoted to work has been unchanged between the end-points of the period, then the increase in the amount of energy available for work contributed about 0.23 percentage points per annum to the annual growth rate of per-capita income ($1.56^{0.0053} - 1 = 0.0023$).

Between 1780 and 1979 British per-capita income grew at an annual rate of about 1.15 per cent (Maddison 1982; Crafts 1985). This bringing the ultra-poor into the labour force and raising the energy available for work by those in the labour force jointly, explain about 30 per cent of the British growth in per-capita income over the past two centuries.

At the present stage of research, the last figure should be considered more illustrative than substantive, since it rests on two implicit assumptions that have yet to be explored adequately. The first is that the share of energy above maintenance allocated to work was the same in 1980 as in *c*.1790. It is difficult to measure the extent or even the net direction of the bias due to this assumption. On the one hand, absenteeism appears to have been much more frequent in the past than at present, due either to poor health or to a lack of labour discipline (Landes 1969). On the other hand, work weeks are shorter today than in the past and a large share of energy above maintenance may be devoted to recreation or other activities whose values are excluded from the national income accounts.[13]

To sum up this section: populations of poor countries, in the past and today, have adapted to their low levels of average food consumption by keeping their body sizes low, by having a large proportion of the population that is excluded from effective labour, and by limiting the intensity of the labour of those who are effectively participating in the labour force. Although this is a feasible way of adjusting to the pressure of population on the food supply, it is a costly solution. It not only impedes economic growth but it undermines health by making individuals more vulnerable to killer infectious diseases, by greatly increasing the rates of chronic diseases among those who survive the acute infections, and by reducing life expectations at all stages of adult ages (cf. Sen 1993). Some may question the use of the experiences of the OECD nations, even if those experiences include periods when these nations were as poor as the poor nations of the Third World, on the ground that the NCHS growth standards do not apply. While that issue needs to be pursued further, it is important to note that recent studies of the growth patterns of affluent Indian children in seven different cities produced growth curves that conformed quite closely to the NCHS standard (Agarwal *et al.* 1991).

3. THE NEW POPULATION ISSUES

Between 1900 and 1950 US life expectation at birth increased from forty-seven to sixty-eight years. Then for the next two decades further progress in longevity came to a virtual halt. Although there were some minor changes in the age-specific death rates of both men and women, they had little impact on the joint expectation of life at birth. During this interregnum a number of thoughtful analysts reviewed the progress in mortality over the preceding century, pointing out why the century-long decline in mortality rates was unique and could not be repeated: nearly all the gains that could be made from the elimination of death from infectious diseases below age 60 had been made. Short of a dramatic breakthrough biologically, it was doubtful that declines in mortality rates at old ages could be as large as those that had already transpired. Indeed, by the early 1960s there was some evidence of a relative deterioration in the mortality rates of persons aged 40 to 70 in a study of Western countries, which appeared to be due to the effects of smoking (Stolnitz 1956-7; Preston 1970; Keyfitz 1977).

It was not until the end of the 1970s that demographers became aware that a new decline in mortality was under way, concentrated this time at older ages. Evidence of a down turn in the death rates of the elderly was contained in medicare data which showed that, beginning in 1968 and continuing through to the end of the 1970s, mortality rates at age 65 and over were declining by 2 per cent per year, and the most rapid advances were concentrated among those aged 85 and older. This development was so unexpected by demographers and epidemiologists that it set off intense discussions, akin to those stimulated two decades earlier as population specialists became aware of the baby boom (Wilkin 1981). The new round of research focused not only on the explanation for the improvement in mortality rates but on how long the decline might continue and whether an increase in the burden of chronic disease was a necessary consequence of the increase in life expectation at older ages (Verbrugge 1984, 1989; Wilson and Drury 1984; Svanborg 1988; Guralnik *et al.* 1989; Riley 1989, 1990; Rothenberg and Koplan 1990).

The most far-reaching aspect of the discussion was the new debate over whether or not the life-span (the ultimate length of the life of a species) is fixed, and if so how long the human life-span was. In a celebrated paper, Fries (1980) reasserted the prevailing gerontological view that, although life expectation had increased from 47 to 73 during the twentieth century, the life-span was fixed. On the assumption that the Gompertz curve (which relates the log of $_nQ_x$ to age) was linear at adult ages, he estimated that the life-span was fixed at 85±7 years. Consequently, age 85 was the upper limit of life expectation and it would be achieved by a rectangularization of the survivorship (l_x) curve: virtually all deaths in a cohort would be compressed into a few years in the neighbourhood of age 85. He also argued that movement of life expectation towards the ideal life-span (85) would not increase the proportion of the elderly population that was disabled, because the onset of chronic diseases could be postponed (the morbidity curve would also be rectangularized) through changes in life-style and biomedical interventions (cf. Olshansky *et al.* 1991).

That paper touched off a highly productive debate which is still in progress. One important aspect of Fries' argument was that chronic diseases had not only replaced acute infectious diseases as the principal medical problem, but that these chronic conditions were independent of the acute infectious diseases. They were instead 'problems of accelerated loss of organ reserve' (p. 132), part of the natural process of senescence that preceded mortality. Since this upper limit of life was fixed at age 85±7 years, the most that could be accomplished by life-style changes and medical intervention was the compression of morbidity against the rigid ceiling of 85±7 (cf. Fries 1983, 1989).

Although the issues raised by Fries have not yet been resolved, much of the evidence accumulated by investigators during the last decade militates against the notion of a genetically fixed life-span or, if it is fixed, suggests that the upper limit is well above 85. Vaupel's study of Danish twins indicates that genetic factors account for only about 30 per cent of the variance in age of death (Vaupel 1991*a*). His study of Swedish males who lived to age 90 indicates that the death rate at that age has declined at a rate of about 1 per cent per annum since 1950, a finding that is contradictory to the rectangularization of the survivorship curve (Vaupel 1991*b*). Two recent studies of insect populations (Carey *et al.* 1992; Curtsinger *et al.* 1992) indicated that variation in environmental conditions had a much larger effect on the life-span than genetic factors, and revealed no pattern suggestive of a fixed upper limit. Collectively, these studies do not rule out genetic factors but suggest something much less rigid than the genetic programming of absolute life-spans: an emerging theory combines genetic susceptibility of various organs with cumulative insults as a result of exposure to risk.

Recent studies indicate that age-specific rates of chronic conditions above age 65 are generally falling, but they do not support the proposition that the span of years during which individuals will be afflicted by chronic diseases is being compressed. According to Manton, Corder, and Stallard (1993), the rate of disability among the elderly in the United States declined by 4.7 per cent between 1982 and 1989. Put on a decade basis, this rate of decline is quite similar to the long-term rates of decline between 1910 and 1985–8 in chronic conditions among elderly veterans (Fogel, Costa, and Kim 1994). The finding is consistent with the growing body of evidence (reported in Section 2 above) indicating that chronic diseases at later ages are, to a considerable degree, the result of exposure to infectious diseases and other types of biomedical and socio-economic stress early in life. It is also consistent with the predicted decline of about 6 per cent per decade in chronic diseases based on the Waaler surface in ill health displayed in Fig. 9.6 (cf. Blair *et al.* 1989; Manton and Soldo 1992; Manton, Stallard, and Singer 1992).

Much of the current research is now focused on explaining the decline in chronic conditions. Part of the emerging explanation is a change in life-styles, particularly the reduction in smoking, the improvement in nutrition, and the increase in exercise, which appear to be involved in reducing the prevalence of coronary heart disease and respiratory diseases. Another part of the explanation is the increasing effectiveness of medical intervention. This point is strikingly demonstrated by comparing the second and last columns of the row on hernias in Table 9.7, above. Before the second World War, hernias, once they occurred, were generally permanent, and often exceedingly painful, conditions. However, by the 1980s about three-quarters of all veterans who had ever had hernias were cured of them. Similar progress over the seven decades is indicated by the row on genito-urinary conditions, which shows that three-quarters of those who had ever had such conditions were cured of them. Other areas where medical intervention has been highly effective include control of hypertension and reduction in the incidence of stroke, surgical removal of osteoarthritis, replacement of knee and hip joints, curing of cataracts, and chemotherapies that reduce the incidence of osteoporosis and heart disease (Manton, Corder, and Stallard 1993).[14]

The success in medical interventions combined with rising incomes has naturally led to a huge increase in the demand for medical services. Econometric estimates suggest a long-run income elasticity in the demand for medical services across OECD nations in the neighbourhood of 1.5, and indicate that 90 per cent of the variance in medical expenditures across OECD countries is explained by variations in income (Moore, Newman, and Fheili 1992). The rapidly growing level of demand, combined with the egalitarian policy of providing medical care at highly subsidized prices, has created the crisis in health-care costs that is now such a focus of public-policy debates across OECD nations, with various combinations of price and governmental rationing under consideration (Schwartz and Aaron 1991; *The Economist* 1991; Newhouse 1992; Schieber, Poullier, and Greenwald 1993).

Whatever the eventual outcome of these policy debates, it is clear that we are in a very different world from that of Malthus. Instead of debating whether to provide food for paupers who might otherwise die, we are now debating how to distribute services that have proved successful in raising the quality of life of the aged and in extending life expectation. And we are now struggling with such entirely new ethical issues as whether it is right to restrict medical services that extend life of a low quality (Wolfe 1986; Shuttleworth 1990; Pellegrino 1993).

Growing opportunity to improve health at young ages, to reduce the incidence of chronic diseases at late ages, and to cure or alleviate the disabilities associated with chronic diseases raises two other post-Malthusian population issues. One is the impact of improved health on population size. A recent paper by Ahlburg and Vaupel (1990) pointed out that, if mortality rates at older ages continue to decline at 2 per cent per annum, the US elderly population of 2050 would be 36 million larger than forecast by the Census Bureau. That possibility poses policy issues with respect to health-care costs (because total medical costs may rise sharply even if cure rates continue to improve) and to pension costs (because the number of persons eligible for benefits under present proposed rules and of projected levels of compensation will become so large that outpayments will exceed planned reserves).

Some policy-makers have sought to meet the pension problem by delaying retirement. Such schemes are based on the proposition that improved health will make it possible for more people to work past age 65. However, the recent findings on the secular improvement in health at older ages make it clear that worsening health is not the explanation for the steep decline since 1890 in labour-force participation rates of males over 65.[15] As Costa (1993) has reported, the US decline in participation rates of the elderly over the past century is largely explained by the rise in income and a decline in the income elasticity of retirement. It is also related to the vast increase in the supply and the quality of leisure-time activities for the labouring classes.

In Malthus's time, and down to the opening of this century, leisure was in very short supply in the OECD countries and, as Veblen (1899) pointed out, it was conspicuously consumed by a small upper class. The typical person laboured over sixty hours per week for wages and often had chores at home which consumed an additional ten or twelve hours. Aside from sleep, eating, and hygiene, such workers usually had barely two hours a day for leisure. Although opera, theatre, and ballet were available, they were too expensive to be consumed by the labouring classes.

Over the twentieth century, hours of work have fallen by nearly half for typical workers. Ironically, those in the top decile of the income distribution have not shared much in this gain of leisure, since the highly paid professionals and business men who populate the top decile work closer to the nineteenth-century standard of 3,200 hours per year than the current working-class standard of about 1,800 hours. There has also been a vast increase in the supply of leisure-time activities—movies, radio, television, amusement parks, participant and spectator sports, travel—and a decline in the price of such activities. Many firms cater especially to the tastes of the elderly, offering reduced prices and special opportunities. As a result, the typical worker spends two-thirds as much time in leisure activities as in work and looks forward to retirement (Fogel 1992a, 1993a).

Given the growing and income-inelastic demand for leisure that characterizes the post-Malthusian milieu of the OECD nations, it remains to be seen to what extent the demand for leisure and retirement can be throttled. Policy-makers may encounter as much resistance to efforts to reduce the implicit subsidies for leisure as they have had recently in raising the taxes on work.

The escape from hunger, poor health, and premature death which began in Malthus's age has not yet run its full course in either the Third World or the OECD world. The Third World, which has reduced mortality rates more than twice as fast as the rich nations did, is still seeking to catch up to current OECD levels (Kelley 1988). However, even the rich nations do not appear to have exhausted the potential for rapid improvement. Studies of particularly long-lived sub-populations (Manton, Stallard, and Tolley 1991) suggest that, within the present state of medical technology, such changes in life-style as the elimination of substance abuse (tobacco, alcohol, and narcotics), improved nutrition, and proper regimens of exercise can, by themselves (without new medical interventions), increase life expectation by fifteen to twenty years. Since more global changes in environment and more far-reaching medical interventions are not included in these estimates, it may well be that the twenty-first century will witness advances in life expectations, reductions in morbidity rates (this time from chronic diseases), and increases in labour productivity that rival those of the twentieth century.

We are on the verge of entering a post-Malthusian world in which not natural resources, but only the rate of technological progress, the supply of key scientific and technical skills, the preference for goods and services over leisure, the prevalence of the egalitarian ethic, and the felicity of political systems will constrain human opportunity. How smooth the transition will be and how rapid the entry of the nations of the Third World will be depends, in no small measure, on the felicity of public policy in the OECD nations.

NOTES

1. The material in the first eight paragraphs of this section draws on a previous paper. I have cited here only a few of the studies on which this compressed survey is based. For more complete documentation, readers should refer to Fogel (1991).
2. Although the English series published by Wrigley and Schofield extends back to 1541, the French series published by the Institut National d'Études Démographiques (INED) currently extends back only to 1740 (INED 1977; Wrigley and Schofield 1981). However, Rebaudo (1979) and Dupâquier (1989) have recently drawn together some samples which, though smaller than

INED samples for the years after 1740, make it possible to compare England and France over the period 1675–1725.

3. From comments made at the Bellagio Conference on Hunger and History, June 1982.

4. The factor 1.27 is obtained on the assumption that eight hours are spent at sleep or at least in bed at a factor of 1.0 BMR, and that essential minimal movements during waking hours have a factor of 1.4 BMR. Hence, over twenty-four hours the factor is $0.6667 * 1.4 + 0.3333 \times 1.0 = 1.27$.

5. Although Toutain's estimate of the national food balance sheet for *c*.1785 may be too low by as much as 200 kcal per capita, the difference between food available for consumption at the household level and national food balance sheets is of the same order of magnitude due to distributional losses. Hence, use of Toutain's estimate for comparison with English estimates based on household purchases is acceptable. For a further discussion of this point see Fogel (1993b).

6. Among the major forces that might change the shape of Waaler surfaces over time are shifts in the c.d.r. *per se* and changes in medical technology. Even a long-term shift in the c.d.r. level, which might be mainly attributable to improved nutrition, in itself need not necessarily alter the Waaler surface (this needs to be further investigated), since the risk levels on a Waaler surface are measured relative to the c.d.r. of the population.

Changes in medical technology are more likely to alter the Waaler surface. For instance, if rapid advances were made in treatments of diseases associated with underweight people but little progress is made in diseases associated with obesity, the relative risk level would fall for low BMIs and rise for high BMIs, tilting the Waaler surface downward on the left. On Fig 9.4 this would show up as a shift of all contour curves to the left, with the distance between contour curves widening in low-BMI regions and vice versa for high-BMI regions.

Changes in life-styles and habits would also change the Waaler surface. A population that smokes would have a different surface from a non-smoking population (low BMIs would have higher risks in a smoking population).

I am indebted to John M. Kim for this note, which is based on work stemming from his dissertation, Kim (1993).

7. I include the aristocracy and other members of the governing classes in the non-agricultural sector even though their wealth was mainly in land, since they were not engaged in farming.

8. Other assumed distributions of the meagre supply of food to the non-agriculture sector yield some output. If the richest 30% seized all of the food, they could have produced about 100 million kcal of work. But then the other 70% of the non-agricultural population would have starved to death in weeks. See

Fogel (1991) for further details, including a discussion of more likely adjustments, such as a drastic reduction in the size of the non-agricultural sector.

9. For recent restatements of the Malthusian theory that focus on fertility, see Schultz (1981) and Von Tunzelman (1986); cf. United Nations (1973), ch. 3. See Easterlin (1968 and 1980) on the baby-boom and cohort-size theories. The results of the European fertility project are summarized in Coale and Watkins (1986). The findings of the World Fertility Survey are reported in Cleland and Scott (1987). See Becker (1991) for an up-to-date statement of the theories of the new household economics and Goldin (1990) for an example of how these theories are applied empirically.

10. In this connection it is worth noting that the rise of the factory reduced the pressure on the food supply by reducing the amount of dietary energy required per hour of labour. This change was due partly to the fact that jobs in light manufacturing industry were less intensive in human energy than farming and partly to the substitution of water-power and mineral energy for dietary energy.

11. I define premature mortality during the eighteenth century as the difference between the average crude death rate of that century and the English death rate of 1980 standardized for the English age structure of 1701–5. The average premature mortality was 21.12 per thousand. Of that amount, 0.83 per thousand was due to crisis mortality, 1.36 was due to the excess mortality of the persons in the next to lowest decile of the English caloric distribution (shown in Table 9.2, above), and 2.16 was due to the excess mortality of those in the lowest decile of the caloric distribution (Fogel 1992*b*; table 9.1). Excess mortality in the two bottom deciles is defined as the difference between the mortality rates of persons of average stature and BMI in England *c*.1790 and the mortality rate of persons with the stature and BMI of persons in the two bottom deciles. The distributions of stature and BMI by caloric deciles are from Fogel and Floud (1994). The relative mortality rates by stature and BMI are from Fogel (1993*b*; table A2).

12. It was assumed that the bottom 20% of the English consuming units were 157 cm tall, with a weight of about 47 kg, which implies a BMR of about 1,342 kcal and about 1,704 kcal for maintenance. The estimated average caloric intake of the second decile of the English caloric distribution in 18–IV was 1,903 kcal. Strolling requires about 76.56 kcal above maintenance, so that such an individual could stroll for about 2.6 hours. One hour of heavy manual labour, including rest breaks, requires 219 kcal above maintenance, and 335 kcal above maintenance while engaged. Computed from the requirements in FAO/WHO/UNU (1985: 76).

13. Although it is my guess that these two influences tend to cancel each other out, it may be that the share of energy above maintenance allocated to work

(measured GNP) is lower now than in the past. In that event the estimate of the share of British economic growth accounted for by improved nutrition and health would be overstated. The other implicit assumption is that the efficiency with which tall people convert energy into work output is the same as that of short people. An enormous literature has developed on this question, but the evidence amassed so far is inconclusive. However, even if both of these assumptions tend to bias upwards the share of British economic growth attributed to improved nutrition, it is quite unlikely that the bias could be as much as 50%. Hence it appears that improved nutrition and health accounted for at least 20% of British economic growth and the best estimate could be as high as 30%.

Of course, there are biases that run in the opposite direction. As Kim has pointed out: 'Depending on how the caloric requirement for BMR and basic maintenance are defined and estimated, it is possible that the actual contribution of improved nutrition and health might be greater than the estimated 30 %. Provided that changes in height and BMI affect not only mortality but also morbidity, shorter- and lower-BMI people will have a higher incidence of disease and illness, which would increase the caloric claims against diet, leaving less calories available for work and also leading to a higher number of sick days. If BMR and basic maintenance fail to take full account of such greater caloric demands by the higher incidence of diseases and illness in a shorter and lighter population, estimates of the effect of improved nutrition and health on economic growth will be biased downward. The shorter and lighter British population of 1790 would have had a higher incidence of disease and illness than the 1980 population, requiring that a greater (negative) adjustment be made to the estimated calories available for work. This leads to a higher estimate of the increase in calories available for work between 1790 and 1980 and hence the contribution of improved health and nutrition would be greater than the estimated 30 %. From a memorandum by John M. Kim dated 4 Nov. 1991.

14. There are, in addition, the vaccines, such as those for polio, rubella, and diphtheria, which have reduced the incidence of once-common diseases and also their costly sequalia, and the antibiotic drugs that have proved so effective in fighting bacterial infections. To the extent that these drugs have reduced biomedical stress at early ages, they may have a significant impact on the reduction of chronic diseases later in the life cycle.

15. Nor can the decline in the participation rate be attributed to the decline in the share of farmers in the labour force, since the retirement rates of farmers were as high as those of non-farmers (Costa 1993).

REFERENCES

Agarwal, K. N., *et al.* (1991), *Growth Performance of Affluent Indian Children (Under-Fives): Growth Standard for Indian Children* (Nutrition Foundation of India, Scientific Report no. 11; New Delhi).

Ahlburg, D. A., and Vaupel, J. W. (1990), 'Alternative Projections of the U.S. Population', *Demography*, 27: 639–52.

Aitchison, J., and Brown, J. A. C. (1966), *The Lognormal Distribution* (University of Cambridge, Department of Applied Economics, Monograph no. 5; Cambridge).

Allen, R. C. (1994), 'Agriculture during the Industrial Revolution, 1700–1850', in R. Floud and D. McCloskey (eds.), *The Economic History of Britain since 1700*, 2nd edn. (Cambridge, forthcoming).

Alter, G., and Riley, J. C. (1989), 'Frailty, Sickness and Death: Models of Morbidity and Mortality in Historical Populations', *Population Studies*, 43: 25–46.

Barker, D. J. P. (1991), 'The Intrauterine Environment and Adult Cardiovascular Disease', in *The Childhood Environment and Adult Disease*, Ciba Research Foundation Symposium 156, 3–16. (Chichester, UK).

Barker, D. J. P. (1993), 'Fetal Origins of Coronary Heart Disease', *British Heart Journal*, 69: 195–6.

Becker, G. S. (1991), *A Treatise on the Family*, 2nd edn. (Cambridge, Mass.).

Bengtsson, T., and Ohlsson, R. (1984), 'Population and Economic Fluctuations in Sweden 1749–1914', in T. Bengtsson, G. Fridlizius, and R. Ohlsson (eds.), *Pre-Industrial Population Change* (Stockholm), 277–97.

Bengtsson, T., and Ohlsson, R. (1985), 'Age-Specific Mortality and Short-Term Changes in the Standard of Living: Sweden, 1751–1859, *European Journal of Population*, 1: 309–26.

Billewicz, W. Z., and MacGregor, I. A. (1982), 'A Birth to Maturity Longitudinal Study of Heights and Weights in Two West African (Gambian) Villages, 1951-1975, *Annals of Human Biology*, 9: 309-20.

Blair, S. N., *et al.* (1989), 'Physical Fitness and All-Cause Mortality: A Prospective Study of Healthy Men and Women', *JAMA* 262/17: 2395–401.

Blayo, Y. (1975), 'Mouvement naturel de la population française de 1740 à 1829', *Population*, 30 (special issue): 15–64.

Carey, J. R., Liedo, P., Orozco, D., and Vaupel, J. W. (1992), 'Slowing of Mortality Rates at Older Ages in Large Medfly Cohorts', *Science*, 258: 457–61.

Cavalli-Sforza, L. L., and Bodmer, W. F. (1971), *The Genetics of Human Populations* (San Francisco).

Chartres, J. A. (1985), 'The Marketing of Agricultural Produce', in J. Thirsk (ed.), *The Agrarian History of England and Wales*, vol. 5: *1640–1750*, pt. 2:

Agrarian Change, 406–502 (Cambridge).

Cipolla, C. M. (1980), *Before the Industrial Revolution: European Society and Economy, 1000–1700* (New York).

Cleland, J., and Scott, C. (1987), *The World Fertility Survey: An Assessment* (Oxford).

Coale, A. J., and Watkins, S. C. (1986), *The Decline of Fertility in Europe* (Princeton, NJ).

Costa, D. L. (1993), 'Health, Income, and Retirement: Evidence from Nineteenth Century America', unpublished Ph. D. dissertation, University of Chicago.

Crafts, N. F. R. (1985), *British Economic Growth during the Industrial Revolution* (Oxford).

Curtsinger, J. W., Fukui, H. H., Townsend, D. R., and Vaupel, J. W. (1992), 'Demography of Genotypes: Failure of the Limited Life-Span Paradigm in *Drosphila melanogaster*', *Science*, 258: 461–63.

Damon, A. (1965), 'Stature Increase among Italian-Americans: Environmental, Genetic, or Both?', *American Journal of Physical Anthropology*, 23: 401–8.

Davidson, S., Passmore, R., Brock, J. F., and Truswell, A. S. (1979), *Human Nutrition and Dietetics*, 7th edn. (Edinburgh).

Dupâquier, J. (1989), 'Demographic Crises and Subsistence Crises in France, 1650–1725', in J. Walter and R. Schofield (eds.), *Famine, Disease and the Social Order in Early Modern Society* (Cambridge), 189–99.

Durnin, J., and Passmore, R. (1967), *Energy, Work and Leisure* (London).

Easterlin, R. A. (1968), *Population, Labor Force, and Long Swings in Economic Growth: The American Experience* (New York).

Easterlin, R. A. (1980), *Birth and Fortune: The Impact of Numbers on Personal Welfare* (New York).

Eckstein, Z., Schultz, T. P., and Wolpin, K. I. (1984), 'Short-Run Fluctuations in Fertility and Mortality in Pre-Industrial Sweden', *European Economic Review*, 26: 297–317.

Elo, I. T., and Preston, S. H. (1992), 'Effects of Early-Life Conditions on Adult Mortality: A Review', *Population Index*, 58: 186–212.

Eveleth, P. B., and Tanner, J. M. (1976), *Worldwide Variation in Human Growth* (Cambridge).

FAO (1977): Food and Agricultural Organization, *The Fourth World Food Survey*, FAO Food and Nutrition Series No. 10, FAO Statistics Series no. 11 (Rome).

FAO/WHO/UNU (1985), 'Energy and Protein Requirements', WHO Technical Report Series no. 724 (Geneva).

Flinn, M. W. (1974), 'The Stabilization of Mortality in Pre-Industrial Western Europe', *Journal of European Economic History*, 3: 285–318.

Flinn, M. W. (1981), *The European Demographic System, 1500–1820* (Baltimore).

Floud, R., and Wachter, K. W. (1982), 'Poverty and Physical Stature: Evidence on the Standard of Living of London Boys, 1770–1870', *Social Science History*, 6: 422–52.

Floud, R., and Wachter, K. W., and Gregory, A. (1990), *Height, Health, and History: Nutritional Status in the United Kingdom, 1750–1980* (Cambridge).

Fogel, R. W. (1987), 'Biomedical Approaches to the Estimation and Interpretation of Secular Trends in Equity, Morbidity, and Labor Productivity in Europe', unpublished paper.

Fogel, R. W. (1989), *Without Consent or Contract: The Rise and Fall of American Slavery*, i (New York).

Fogel, R. W. (1991), 'New Findings on Secular Trends in Nutrition and Mortality: Some Implications for Population Theory', unpublished paper.

Fogel, R. W. (1992a), 'Egalitarianism: The Economic Revolution of the Twentieth Century', The 1992 Simon Kuznets Memorial Lectures Presented at Yale University (22–4 April).

Fogel, R. W. (1992b), 'Second Thoughts on the European Escape from Hunger: Famines, Chronic Malnutrition, and Mortality', in S. R. Osmani (ed.), *Nutrition and Poverty* (Oxford), 243–86.

Fogel, R. W. (1993a), 'A Comparison of Biomedical and Economic Measures of Egalitarianism: Some Implications of Secular Trends for Current Policy', paper presented at the Workshop on Economic Theories of Inequality, Stanford University (11–13 March).

Fogel, R. W. (1993b), 'New Sources and New Techniques for the Study of Secular Trends in Nutritional Status, Health, Mortality, and the Process of Aging', *Historical Methods*, 26: 5–43.

Fogel, R. W. (1994), 'Economic Growth, Population Theory, and Physiology: The Bearing of Long–Term Processes on the Making of Economic Policy', *American Economic Review*, 84: 369–95.

Fogel, R. W. (1995), *The Escape from Hunger and High Mortality: Europe, America, and the Third World 1700–2050*, forthcoming.

Fogel, R. W., Costa, D. L., and Kim, J. M. (1994), 'Secular Trends in the Distribution of Chronic Conditions and Disabilities at Young Adult and Late Ages, 1860–1988: Some Preliminary Findings', paper presented at the NBER Summer Institute, Economics of Aging Program (26–8 July 1993).

Fogel, R. W. , and Floud, R. (1994), 'Nutrition and Mortality in France, Britain, and the United States', unpublished paper.

Fogel, R. W. , Galantine, R., and Manning, R. L (1992) (eds.), *Without Consent or Contract*, ii. *Evidence and Methods* (New York).

Fogel, R. W. *et al.* (1983), 'Secular Changes in American and British Stature and Nutrition', *Journal of Interdisciplinary History*, 14: 445–81.

Fogel, R. W. *et al.* (1986), 'The Aging of Union Army Men: A Longitudinal Study', unpublished paper.

Freudenberger, H., and Cummins, G. (1976), 'Health, Work, and Leisure Before the Industrial Revolution', *Explorations in Economic History*, 13: 1-12.

Fridlizius, G. (1984), 'The Mortality Decline in the First Phase of the Demographic Transition: Swedish Experiences', in T. Bengtsson, G. Fridlizius, and R. Ohlsson (eds.), *Pre-Industrial Population Change* (Stockholm), 71-114.

Friedman, G. C. (1982), 'The Heights of Slaves in Trinidad', *Social Science History*, 6: 482–515.

Fries, J. F. (1980), 'Ageing, Natural Death, and the Compression of Morbidity', *New England Journal of Medicine*, 303: 130–6.

Fries, J. F. (1983), 'The Compression of Morbidity', *Milbank Quarterly*, 61: 397–419.

Fries, J. F. (1989), 'The Compression of Morbidity: Near or Far?', *Milbank Quarterly*, 67: 208–32.

Galloway, P. (1986), 'Differentials in Demographic Responses to Annual Price Variations in Pre-Revolutionary France: A Comparison of Rich and Poor Areas in Rouen, 1681–1787', *European Journal of Population*, 2: 269–305.

Galloway, P. (1987), 'Population, Prices and Weather in Preindustrial Europe', Ph.D. dissertation, University of California, Berkeley.

Goldin, C. D. (1990), *Understanding the Gender Gap: An Economic History of American Women* (New York).

Goubert, P. (1960), *Beauvais et le Beauvaisis de 1600 à 1730* (Paris).

Goubert, P. (1965), 'Recent Theories and Research in French Population between 1500 and 1700', in D. V. Glass and D. E. C. Eversley (eds.), *Population in History: Essays in Historical Demography* (Chicago), 457–73.

Goubert, P. (1973), *The Ancien Régime*, trans. S. Cox (New York).

Gould, B. A. (1869), *Investigations in the Military and Anthropological Statistics of American Soldiers* (Cambridge, Mass.).

Grantham, G. (1990), 'Some Implications of the Level of Agricultural Labour Productivity in Eighteenth and Early Nineteenth Century France', paper presented at the Annual Conference of the Agricultural History Association, April, Leeds.

Guralnik, J. M., LaCroix, A. Z., Everett, D. F., and Kovar, M. G. (1989), 'Aging in the Eighties: The Prevalence of Comorbidity and its Association with Disability', *Advance Data*, No. 170, 26 May.

Heywood, P. F. (1983), 'Growth and Nutrition in Papua New Guinea', *Journal of Human Evolution*, 12: 131–43.

Himmelfarb, G. (1983), *The Idea of Poverty: England in the Early Industrial Age* (New York).

Holderness, B. A. (1989), 'Prices, Productivity, and Output', in J. Thirsk (ed.),

The Agrarian History of England and Wales, G. E. Mingay (ed.), vi. *1750–1850* (Cambridge), 84–189.

Horton, S. (1984), 'Nutritional Status and Living Standards Measurement', mimeo., World Bank.

INED (1977): Institut National d'Études Démographiques, 'Sixième rapport sur la situation démographique de la France', *Population*, 32: 253–338.

Kelley, A. C. (1988), 'Economic Consequences of Population Change in the Third World', *Journal of Economic Literature*, 26: 1685–728.

Keyfitz, N. (1977), 'What Difference would it Make if Cancer were Eradicated? An Examination of the Taeuber Paradox', *Demography*, 14: 411–18.

Kim, J. M. (1993), 'Waaler Surfaces: A New Perspective on Height, Weight, Morbidity, and Mortality', unpublished paper.

Komlos, J. (1990), 'Stature, Nutrition and the Economy in the Eighteenth-Century Habsburg Monarchy', unpublished Ph.D. dissertation, University of Chicago.

Landes, D. S. (1969), *The Unbound Prometheus* (Cambridge).

Laslett, P. (1965), *The World we have Lost: England before the Industrial Age*, 3rd edn. (New York, 1984).

Lebrun, F. (1971), *Les Hommes et la mort en Anjou aux 17ᵉ et 18ᵉ siècles* (Paris).

Lee, R. (1981), 'Short-Term Variation: Vital Rates, Prices, and Weather', in E. A. Wrigley and R. S. Schofield (eds.), *The Population History of England, 1541–1871: A Reconstruction*, (Cambridge, Mass.) 356–401.

Lindert, P. H., and Williamson, J. G. (1982), 'Revising England's Social Tables: 1688-1812', *Explorations in Economic History*, 19: 385-408.

Lindert, P. H., and Williamson, J. G. (1983), 'Reinterpreting Britain's Social Tables, 1688–1913', *Explorations in Economic History*, 20: 94–109.

Lipton, M. (1983), 'Poverty, Undernutrition and Hunger', World Bank, Staff Working Papers No. 597.

Livi-Bacci, M. (1990), *Population and Nutrition: An Essay on European Demographic History* (New York).

Livi-Bacci, M. (1992), *A Concise History of World Population* (Cambridge).

Mackenbach, J. P. (1992), 'Socio-Economic Health Differences in the Netherlands: A Review of Recent Empirical Findings', *Social Science and Medicine*, 34: 213–26.

McKeown, T. (1976), *The Modern Rise of Population* (New York).

McKeown, T. (1978) 'Fertility, Mortality and Cause of Death: An Examination of Issues Related to the Modern Rise of Population', *Population Studies*, 32: 535–42.

McMahon, M. M., and Bistrian, B. R. (1990), 'The Physiology of Nutritional Assessment and Therapy in Protein-Calorie Malnutrition', *Disease-a-Month*, 36: 373–417.

Maddison, A. (1982), *Phases of Capitalist Development* (Oxford).

Malthus, T. R. (1798), *An Essay on the Principle of Population*, ed. Anthony Flew (Harmondsworth, 1976).

Manton, K. G., and Soldo, B. J. (1992), 'Disability and Mortality among the Oldest Old: Implications for Current and Future Health and Long-Term Care Service Needs', in R. M. Suzman, D. P. Willis, and K. G. Manton (eds.), *The Oldest Old* (Oxford), 199–250.

Manton, K. G., Corder, L. S., and Stallard, E. (1993), 'Estimates of Change in Chronic Disability and Institutional Incidence and Prevalence Rates in the U.S. Elderly Population from the 1982, 1984, and 1989 National Long-Term Care Survey', photocopy, Duke University, Center for Demographic Studies.

Manton, K. G., Stallard, E., and Tolley, H. D. (1991), 'Limits to Human Life Expectancy: Evidence, Prospects, and Implications', *Population and Development Review*, 17: 603–37.

Manton, K. G., Stallard, E., and Singer, B. (1992), 'Projecting the Future Size and Health Status of the U.S. Elderly Population', *International Journal of Forecasting*, 8: 433–58.

Martorell, R. (1985), 'Child Growth Retardation: A Discussion of its Causes and its Relationship to Health', in K. Blaxter and J. C. Waterlow (eds.), *Nutritional Adaptation in Man* (London), 13–29.

Martorell, R. (1990), 'Consequences of Stunting in Early Childhood for Adult Body Size in Rural Guatemala', *Annales Nestlé*, 48: 85–92.

Meerton, M. A. von. (1989), 'Croissance économique en France et accroissement des français: Une analyse "Villermetrique"', typescript, Center voor Economische Studiën, Leuven.

Meuvret, J. (1946), 'Les Crises de subsistances et la demographie de la France d'Ancien Régime', *Population*, 1: 643–50.

Moore, W. J., Newman, R. J., and Fheili, M. (1992), 'Measuring the Relationship between Income and NHEs', *Health Care Financing Review*, 14: 133–39.

Mueller, W. H. (1986), 'The Genetics of Size and Shape in Children and Adults', in F. Falkner and J. M. Tanner (eds.), *Human Growth*, 2nd edn., iii., *Methodology* (New York), 145–68.

Newhouse, J. P. (1992), 'Medical Care Costs: How Much Welfare Loss?', *Journal of Economic Perspectives*, 6/3: 3–21.

O'Brien, P., and Keyder, C. (1978), *Economic Growth in Britain and France 1780-1914: Two Paths to the Twentieth Century* (London).

Oddy, D. J. (1990), 'Food, Drink and Nutrition', in F. M. L. Thompson (ed.), *The Cambridge Social History of Britain 1750–1950*, ii. *People and their Environment* (New York), 251–78.

Olshansky, S. J., Rudberg, M. A., Carnes, B. A., Cassel, C. K., and Brody, J. A. (1991), 'Trading Off Longer Life for Worsening Health: The Expansion of

Morbidity Hypothesis', *Journal of Aging and Health*, 3/2: 194–216.

Osmani, S. R. (1992), 'On Some Controversies in the Measurement of Undernutrition', in S. R. Osmani (ed.), *Nutrition and Poverty* (Oxford), 121–64.

Pellegrino, E. D. (1993), 'The Metamorphosis of Medical Ethics: A 30-Year Retrospective', *Journal of the American Medical Association*, 269: 1158–62.

Perrenoud, A. (1984), 'The Mortality Decline in a Long-Term Perspective', in T. Bengtsson, G. Fridlizius, and R. Ohlsson (eds.), *Pre-Industrial Population Change* (Stockholm), 41–69.

Perrenoud, A. (1991), 'The Attenuation of Mortality Crises and the Decline of Mortality', in R. Schofield, D. Reher, and A. Bideau (eds.), *The Decline of Morality* (Oxford), 18–37.

Preston, S. H. (1970), 'An International Comparison of Excessive Adult Mortality', *Population Studies*, 24: 5–20.

Preston, S. H. (1976), *Mortality Patterns in National Populations* (New York).

Preston, S. H., and van de Walle, E. (1978), 'Urban French Mortality in the Nineteenth Century', *Population Studies*, 32: 275–97.

Quenouille, M. H., Boyne, A. W., Fisher, W. B., and Leitch, I. (1951), 'Statistical Studies of Recorded Energy Expediture in Man', Technical Communication No. 17, Commonwealth Bureau of Animal Nutrition.

Rebaudo, D. (1979), 'Le Mouvement annuel de la population française rurale de 1670 à 1740', *Population*, 34: 589–606.

Richards, T. (1984), 'Weather, Nutrition and the Economy: The Analysis of Short Run Fluctuations in Births, Deaths and Marriages, France 1740–1909', in T. Bengtsson, G. Fridlizius, and R. Ohlsson (eds.), *Pre-Industrial Population Change* (Stockholm), 357–89.

Riley, J. C. (1989), *Sickness, Recovery and Death: A History and Forecast of Ill Health* (Iowa City).

Riley, J. C. (1990), 'Long-Term Morbidity and Mortality Trends: Inverse Health Transitions', in J. Caldwell *et al.* (eds.), *What We Know about Health Transition: The Cultural, Social and Behavioral Determinants of Health; The Proceedings of an International Workshop, Canberra, May 1989* (Canberra) i. 165–87 .

Robbins, S. L., Cotran, R. S., and Kumar, V. (1984), *Pathologic Basis of Disease*, 3rd edn. (Philadelphia. Pa).

Rothenberg, R. B., and Koplan, J. P. (1990), 'Chronic Disease in the 1990s', *Annual Review of Public Health*, 11: 267–96.

Schieber, G. J., Poullier, J.-P., and Greenwald, L. M. (1993), 'Health Care Systems in Twenty-Four Countries', *Health Affairs*, 10/3: 22–38.

Schofield, R., and Reher, D. (1991), 'The Decline of Mortality in Europe', in R. Schofield, D. Reher, and A. Bideau (eds.), *The Decline of Mortality in Europe* (Oxford), 1–17.

Schultz, T. P. (1981), *Economics of Population* (Reading, Mass.).

Schwartz, W. B., and Aaron, H. J. (1991), 'Must We Ration Health Care?', *Best's Review*, Jan.: 37–41.

Scrimshaw, N. S., Taylor, C. E., and Gordon, J. E. (1968), *Interactions of Nutrition and Infection* (Geneva).

Seckler, D. (1980), ' "Malnutrition": An Intellectual Odyssey', *Western Journal of Agricultural Economics*, 5: 219–27.

Seckler, D. (1982), 'Small but Healthy? A Basic Hypothesis in the Theory, Measurement and Policy of Malnutrition', in P. V. Sukhatme (ed.), *Newer Concepts in Nutrition and Their Implications for Policy* (Pune).

Sen, A. (1993), 'The Economics of Life and Death', *Scientific American*, May: 40–7.

Shammas, C. (1990), *The Pre-Industrial Consumer in England and America* (Oxford).

Shuttleworth, J. S. (1990), 'Ethical Issues of Cost in Long-Term Care', *Journal of the Medical Association of Georgia*, 79: 843–5.

Smith, D. S. (1977), 'A Homeostatic Demographic Regime: Patterns in West European Family Reconstruction Studies', in R. D. Lee (ed.), *Population Patterns in the Past* (New York), 19–51.

Steckel, R. (1986), 'A Peculiar Population: The Nutrition, Health, and Mortality of American Slaves from Childhood to Maturity', *Journal of Economic History*, 46: 721–41.

Steckel, R. (1987), 'Growth Depression and Recovery: The Remarkable Case of American Slaves', *Annals of Human Biology*, 14: 101–10.

Stigler, G. (1954), 'The Early History of Empirical Studies of Consumer Behavior', *Journal of Political Economy*, 62: 95–113.

Stolnitz, G. (1956–7), 'A Century of International Mortality Trends: II', *Population Studies*, 10: 17–42.

Sukhatme, P. (1981), *Relationship between Malnutrition and Poverty*, Indian Association of Social Science Institutions, First National Conference on Social Sciences, Delhi, 12-15 January 1981.

Svanborg, A. (1988), 'The Health of the Elderly Population: Results from Longitudinal Studies with Age-Cohort Comparisons', in *Research and the Aging Population*, Ciba Research Foundation Symposium 134 (Chichester, UK), 3–16.

Tanner, J. M. (1982) 'The Potential of Auxological Data for Monitoring Economic and Social Well-Being', *Social Science History*, 6: 571–81.

The Economist (1990), 'Squeezing in the Next Five Billion', 20 Jan.: 19–20, 22.

The Economist (1991), 'A Spreading Sickness', 6 July: S1–S18.

Thompson, E. P. (1967), 'Time, Work-Discipline, and Industrial Capitalism', *Past and Present*, 38: 57–97.

Toutain, J. (1971), 'La Consommation alimentaire en France de 1789 à 1964', *Economies et sociétés, Cahiers de l'I.S.E.A.*, 5: 1909-2049.

United Nations (1953), *The Determinants and Consequences of Population Trends* (Population Studies No. 17; New York).

United Nations (1973), *The Determinants and Causes of Population Trends* (Population Studies No. 50; New York).

United Nations (1992), *Second Report on the World Nutrition Situation*, i.: *Global and Regional Results* (Geneva).

U.S. Department of Health and Human Services (1987), 'Anthropometric Reference Data and Prevalence of Overweight', *Vital and Health Statistics*, ser. 11, no. 238 (Washington, DC).

U.S. National Center for Health Statistics (1977), 'Dietary Intake Findings: United States, 1971–1974', Data from the Health and Nutrition Examination Survey, Health Resources Administration, Public Health Service, US Department of Health, Education and Welfare, Series 11, no. 202.

Van Wieringen, J. C. (1978), 'Secular Growth Changes', in F. Falkner and J. M. Tanner (eds.), *Human growth*, ii. *Postnatal growth*, 445–72. (New York).

Van Wieringen, J. C. (1986), 'Secular Growth Changes', in F. Falkner and J. M. Tanner, *Human growth*, 2nd edn., iii. *Methodology* (New York), 307–71.

Vaupel, J. W. (1991*a*), 'The Impact of Population Aging on Health and Health Care Costs: Uncertainties and New Evidence about Life Expectancy', unpublished paper.

Vaupel, J. W. (1991*b*), 'Prospects for a Longer Life Expectancy', paper presented to the annual meeting of the Population Association of America, Washington, DC (21–3 March).

Veblen, T. (1899), *The Theory of the Leisure Class: An Economic Study of Institutions* (New York, 1934).

Verbrugge, L. M. (1984), 'Longer Life but Worsening Health? Trends in Health and Mortality of Middle-Aged and Older Persons', *Health and Society*, 62: 475–519.

Verbrugge, L. M. (1989), 'Recent, Present, and Future Health of American Adults', *Annual Review of Public Health*, 10: 333–61.

Von Tunzelman, G. N. (1986), 'Malthus's "Total Population System": A Dynamic Reinterpretation', in D. Coleman and R. Schofield (eds.), *The State of Population Theory: Forward from Malthus* (Oxford), 65–95.

Waaler, H. T. (1984), 'Height, Weight and Mortality: The Norwegian Experience', *Acta Medica Scandinavica*, supplement. 679: 1–51.

Weir, D. R. (1982), 'Fertility Transition in Rural France, 1740–1829', unpublished Ph. D. dissertation, Stanford University.

Weir, D. R. (1989), 'Markets and Mortality in France, 1600–1789', in J. Walter and R. Schofield (eds.), *Famine, Disease, and the Social Order in Early Mo-*

dern Society (Cambridge), 201–34.

Weir, D. R. (1993), 'Parental Consumption Decisions and Child Health during the Early French Fertility Decline, 1790–1914', *Journal of Economic History*, 53: 259–74.

West, P., Macintyre, S., Annandale, E., and Hunt, K. (1990), 'Social Class and Health in Youth: Findings from The West of Scotland Twenty-07 Study', *Social Science and Medicine*, 30: 665–73.

Wilkin, J. C. (1981), 'Recent Trends in the Mortality of the Aged', *Society of Actuaries Transactions*, 33: 11–62.

Williamson, J. G. (1985), *Did British Capitalism Breed Inequality?* (Boston).

Wilson, R. W., and Drury, T. F. (1984), 'Interpreting Trends in Illness and Disability: Health Statistics and Health Status', *Annual Review of Public Health*, 5: 83–106.

Wolfe, B. L. (1986), 'Health Status and Medical Expenditures: Is There a Link?', *Social Science and Medicine*, 22: 993–99.

World Bank (1987),*World Development Report 1987* (Oxford).

World Bank (1992),*World Development Report 1992* (Oxford).

Wrigley, E. A. (1987), 'Urban Growth and Agricultural Change: England and the Continent in the Early Modern Period', in *People, Cities and Wealth: The Transformation of Traditional Society* (Oxford), 157–93.

Wrigley, E. A. and Schofield, R. S. (1981), *The Population History of England, 1541-1871: A Reconstruction* (Cambridge, Mass.).

POPULATION, ECONOMIC
DEVELOPMENT, AND THE ENVIRONMENT